Game Theory in Communication Networks

Cooperative Resolution of
Interactive Networking Scenarios

OTHER TELECOMMUNICATIONS BOOKS FROM AUERBACH

Ad Hoc Mobile Wireless Networks:
Principles, Protocols and Applications
Subir Kumar Sarkar, T.G. Basavaraju,
and C. Puttamadappa
ISBN 978-1-4200-6221-2

Communication and Networking in Smart Grids
Yang Xiao (Editor), ISBN 978-1-4398-7873-6

Decentralized Control and Filtering in
Interconnected Dynamical Systems
Magdi S. Mahmoud
ISBN 978-1-4398-3814-3

Delay Tolerant Networks: Protocols and
Applications
Athanasios V. Vasilakos, Yan Zhang, and
Thrasyvoulos Spyropoulos
ISBN 978-1-4398-1108-5

Emerging Wireless Networks: Concepts,
Techniques and Applications
Christian Makaya and Samuel Pierre (Editors)
ISBN 978-1-4398-2135-0

Game Theory in Communication Networks:
Cooperative Resolution of Interactive
Networking Scenarios
Josephina Antoniou and Andreas Pitsillides
ISBN 978-1-4398-4808-1

Green Mobile Devices and Networks:
Energy Optimization and Scavenging
Techniques
Hrishikesh Venkataraman and
Gabriel-Miro Muntean (Editors)
ISBN 978-1-4398-5989-6

Handbook on Mobile Ad Hoc and
Pervasive Communications
Laurence T. Yang, Xingang Liu, and
Mieso K. Denko (Editors)
ISBN 978-1-4398-4616-2

IP Telephony Interconnection Reference:
Challenges, Models, and Engineering
Mohamed Boucadair, Isabel Borges,
Pedro Miguel Neves, and Olafur Pall Einarsson
ISBN 978-1-4398-5178-4

Measurement Data Modeling and
Parameter Estimation
Zhengming Wang, Dongyun Yi, Xiaojun Duan,
Jing Yao, and Defeng Gu
ISBN 978-1-4398-5378-8

Media Networks: Architectures,
Applications, and Standards
Hassnaa Moustafa and Sherali Zeadally (Editors)
ISBN 978-1-4398-7728-9

Multimedia Communications and Networking
Mario Marques da Silva, ISBN 978-1-4398-7484-4

Near Field Communications Handbook
Syed A. Ahson and Mohammad Ilyas (Editors)
ISBN 978-1-4200-8814-4

Next-Generation Batteries and Fuel Cells for
Commercial, Military, and Space Applications
A. R. Jha, ISBN 978-1-4398-5066-4

Physical Principles of Wireless
Communications, Second Edition
Victor L. Granatstein, ISBN 978-1-4398-7897-2

Security of Mobile Communications
Noureddine Boudriga, ISBN 978-0-8493-7941-3

Smart Grid Security: An End-to-End View
of Security in the New Electrical Grid
Gilbert N. Sorebo and Michael C. Echols
ISBN 978-1-4398-5587-4

Systems Evaluation: Methods, Models,
and Applications
Sifeng Liu, Naiming Xie, Chaoqing Yuan,
and Zhigeng Fang
ISBN 978-1-4200-8846-5

Transmission Techniques for Emergent
Multicast and Broadcast Systems
Mario Marques da Silva, Americo Correia,
Rui Dinis, Nuno Souto, and Joao Carlos Silva
ISBN 978-1-4398-1593-9

TV Content Analysis: Techniques and
Applications
Yiannis Kompatsiaris, Bernard Merialdo,
and Shiguo Lian (Editors)
ISBN 978-1-4398-5560-7

TV White Space Spectrum Technologies:
Regulations, Standards, and Applications
Rashid Abdelhaleem Saeed and
Stephen J. Shellhammer
ISBN 978-1-4398-4879-1

Wireless Sensor Networks:
Principles and Practice
Fei Hu and Xiaojun Cao
ISBN 978-1-4200-9215-8

AUERBACH PUBLICATIONS
www.auerbach-publications.com
To Order Call: 1-800-272-7737 • Fax: 1-800-374-3401

Game Theory in Communication Networks

Cooperative Resolution of Interactive Networking Scenarios

Josephina Antoniou and Andreas Pitsillides

CRC Press
Taylor & Francis Group
Boca Raton London New York

CRC Press is an imprint of the
Taylor & Francis Group, an **informa** business

CRC Press
Taylor & Francis Group
6000 Broken Sound Parkway NW, Suite 300
Boca Raton, FL 33487-2742

First issued in paperback 2016

© 2013 by Taylor & Francis Group, LLC
CRC Press is an imprint of Taylor & Francis Group, an Informa business

No claim to original U.S. Government works

Version Date: 20120611

ISBN 13: 978-1-138-19938-5 (pbk)
ISBN 13: 978-1-4398-4808-1 (hbk)

Library of Congress Cataloging-in-Publication Data

Antoniou, Josephina.
    Game theory in communication networks : cooperative resolution of interactive networking scenarios / Josephina Antoniou, Andreas Pitsillides.
        p. cm.
    Includes bibliographical references and index.
    ISBN 978-1-4398-4808-1 (hardback)
    1. Telecommunication systems--Computer simulation. 2. Telecommunication systems--Mathematical models. 3. Internetworking (Telecommunication)--Computer simulation. 4. Game theory. I. Pitsillides, A. (Andreas) II. Title.

TK5102.83.A58 2012
621.38201'5193--dc23   2012017888

Visit the Taylor & Francis Web site at
http://www.taylorandfrancis.com

and the CRC Press Web site at
http://www.crcpress.com

Contents

Preface

Communication Networks is an area in which multiple interactive situations arise among networking entities. Such interactive situations may be hard to resolve satisfactorily because of the conflicting goals of the participating entities. It is often the case that cooperation in such interactive situations may be beneficial to all entities involved; however, cooperation is normally hard to enforce, unless specific conditions can be applied. An example of such a network is a converged communication network, a paradigm found in Fourth Generation Telecommunication Networks, where heterogeneous access technologies may coexist. Among other features, this new network model enables a user (or a set of users) to be served by any (one or many) of the multiple, available access networks. These access networks carry differing characteristics and capabilities encouraging the decoupling of carriage and content, i.e., the infrastucture operators and the service or content providers can be different entities in this new system. As part of the same network, these entities may need to cooperate in order to serve the users of the networks while maintaining their own goals as self-managed entities. For this specific network, the common thread that links all this heterogeneity is the support for a user-centric paradigm of communication, converging all activities to the system's key function, i.e., to satisfy its customers. Cooperation in Fourth Generation Communication Networks can take advantage of these varying characteristics, and exploit them in complementary manners in order to surpass any limits imposed by any one of these networks on their own, through appropriate network synergies.

Synergies, i.e., cooperation between participating entities in communication networks, promote the useful coexistence of the participating entities, aiming at enhancing the overall network, since the support of demanding services, as for example interactive and multiparty multimedia services, can become a challenging task due to the heterogeneity of the entities involved, the user(s), and the access network(s). This heterogeneity results in different and often conflicting interests for these entities. Since cooperation between these entities, if achieved, is expected to be beneficial, we explore examples of interactive situations arising in such communication networks, and show how cooperation is beneficial for the interacting entities, i.e., how the proposed cooperative modes of behavior allow the interacting entities to achieve their own satisfaction, despite their conflicting interests, and how this cooperation can be encouraged.

In fact, there exist multiple interactive situations in communication networks among the entities participating in such networks where cooperation of the entities might be a beneficial or even a necessary way to achieve certain goals of the participating entities. A rich theory behind cooperation, often referred to as *cooperation theory*, may be used to analyze and resolve such situations in a cooperative manner. Our aim in this book is to make use of cooperation theory, focusing on game theoretic tools, and targeting specific examples from communication networks to demonstrate how cooperation through the use of such theoretical tools can prove beneficial in resolving situations of interaction.

Interactions between entities with conflicting interests follow action plans designed, by each entity, in such a way as to achieve a particular selfish goal; such interactions are known as strategic interactions. Strategic interactions are studied by Game Theory, a field which develops models that prescribe actions for entities interacting in a strategic manner, such that they achieve satisfactory gains from the situation. *To target the question of how to promote cooperative behavior in interactive situations between heterogeneous entities in communication networking scenarios, the book utilizes game theoretical models to analyze a set of illustrative strategic situations and demonstrates profitable behaviors of the participating entities.*

In the course of this book, it is shown how cooperation can be motivated in each of the selected interactive situations and, furthermore, that such cooperative behavior is beneficial for the interacting entities. Cooperation is motivated by characteristics of the selected situations such as repeatability of interaction, need for sharing between entities, and need for participation in groups.

Our objective is not to offer a comprehensive book on cooperation theory, where a rich bibliography already exists; instead we want to illustrate the nature and power of the theory and its applicability and benefits in interactive situations selected within the area of communication networks by targeting specific examples of such interactive scenarios. The richness of the theory on cooperation may be further explored by the interested reader who wants to expand on specific scenarios of interest not covered in the book, by following on the given bibliography, and beyond.

The structure of the book considers first the presentation of theory that can be used to promote cooperation for the entities in a particular interactive situation. Then, for each chapter, the first part introduces basic theory for dealing with a particular interactive situation in an attempt to show how particular aspects of game theory can be usefully employed to formulate and solve interactive situations commonly appearing in the field of communication networks, and the second part of each chapter presents example scenarios, showing the applicability and power of the theory, from the scenario's view, in order to demonstrate a number of cooperative interactions and discuss how these could be addressed within the theoretical framework presented in the first part of the chapter. In particular, four specific situations are covered in the book. Chapter 2 and chapter 3 deal with two-player interaction, while the next two chapters deal with interactions between multiple players. Finally, chapter 6 deals with a performance evaluation framework based on MATLAB.

Beyond the great pioneers of game theory, who contributed to the general theory of Games (with several Nobel prizes awarded for contributions in this field), several colleagues contributed in realizing the scientific contents presented in this book (including the anonymous reviewers of several published articles by the authors in this area), to whom the authors are very grateful for their cooperation.

Primarily, the authors acknowledge Dr. Vicky Papadopoulou (European University Cyprus) for her valuable guidance in the theoretical resolutions presented in the book as well as for her beneficial revisions of a large part of the book content.

Furthermore, the authors wish to acknowledge Dr. Vasos Vassiliou and Dr. Chris Christodoulou (University of Cyprus), who provided the motivation for the idea of adaptivity as an enhancement to the game theoretical framework presented in the user-network interaction scenario analyzed in Chapter 2. For the same chapter, Dr. Loizos Michael (Open University of Cyprus) has offered very useful comments and suggestions as well.

Moreover, the specifics and unfolding of the scenario presented in Chapter 4 are largely contributed by Dr. Lavy Libman (University of Sydney), whom the authors wish to thank for his valuable input.

Finally, the study of cooperation between multiple networks in Chapter 5 would not have been possible without the help and cooperation of Prof. Ioannis Stavrakakis, Dr. Ioannis Koukoutsidis and Dr. Eva Jaho, having contributed extensively in the theoretical analysis of coalition formation possibilities.

Special thanks also go to the staff of CRC Press, especially Rich O'Hanley and Stephanie Morkert, for their patience and professionalism from the inception of the book until its publication.

Last but not least, we would like to thank our families and friends for their unwavering encouragement and support throughout writing this book.

Chapter 1

Introduction: Game theory as an analytical tool

Game theory is a theoretical framework that attempts to mathematically capture both human and non-human (e.g. computer, animal, plant) behavior during a strategic situation. A strategic situation is a situation that involves the interaction of two or more entities in which the individual's success depends on the choice of actions by others. A logical behavior would be to attempt to find equilibria between the entities (called the *players*), i.e., sets of strategies (action sequences) that players will unlikely want to change, since if they do they will most probably benefit less. Therefore, game theory can be used to model situations of interaction and can offer solutions so that mutually agreeable sequences of actions can be employed by the players; game theoretic models make the assumption that the entities make rational choices, i.e., choices that are profitable according to each entity's own interpretation of profit.

In order for a strategic situation to become a *game* between two or more players, there must be a mutual awareness of the participants regarding the crosseffect of their actions. A strategic situation, where the actions of a participant may alter another's outcome, is primarily characterized by the players' strategies. In addition, a strategic situation contains other elements that must be taken into consideration when modelling such a situation as a game, e.g., chance and skill (elements that are not easily controlled or modified). Game models, i.e., models of specific strategic situations, may be categorized in various ways due to the several elements that they contain. A usual categorization is made by looking at the players' movements; if they are sequential we have an extensive game form (a.k.a. *sequential-moves game* form), whereas if they are simultaneous it is referred to as *normal*. The extensive form can be used to formalize games with a time sequencing of moves. Sequential games are illustrated on *trees*, joined decision trees for all of the players in the game, like the example *tree* in Figure 1.1 illustrating all of the possible actions that can be taken by all of the players and indicating all of the possible outcomes of the game. This book employs the notion of game trees, as in the example depicted in Figure 1.1. Each node represents a point of choice for a player. One player is specified at each node. The links between the nodes represent a possible action for that player. The payoffs are specified at the bottom of the tree.

In Figure 1.1 there are two players. Player 1 moves first and chooses either A or B. Player 2 knows the move chosen by Player 1 and hence chooses C or D. Each sequence of choices from the players results in a different set of payoffs for the two players. Depending on the outcome of a game, a payoff is provided for each player, which reflects their own gain at the end of the game (or each round of the game in the case of an iterative game).

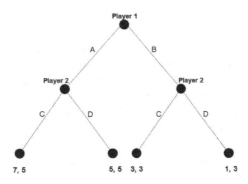

Figure 1.1: An example tree of a sequential form game.

The normal form game (a.k.a. strategic form) is usually represented by a *table* which shows the players, strategies, and payoffs, as the example in Table 1.1. In the example there are two players; one chooses the row and the other chooses the column. Each player has two strategies, which are specified by the number of the row and the number of the column. The payoffs are provided in the interior. Two numbers are provided, one for each player. The first number is the payoff received by the row player (Player 1); the second is the payoff for the column player (Player 2). Suppose that Player 1 plays A and that Player 2 plays C. Then Player 1 gets a payoff of 4, and Player 2 gets 3. When a game is presented in normal form, it is presumed that each player acts simultaneously or, at least, without knowing the actions of the other. Every extensive game form has an equivalent normal game form.

Furthermore, an interaction may happen only once or repeatedly; in the first situation we are faced with *one-shot game models*, while the second situa-

Table 1.1: An example table of a normal game form

	Player 2 plays C	Player 2 plays D
Player 1 plays A	4, 3	2, 1
Player 1 plays B	0, 0	−1, −2

tion requires *repeated game models*. Many of the real-world problems modelled employ game players that interact in finitely many moves (one or more). However, one class of games studied is the one of infinitely repeated games, a model used also for games whose horizon (i.e., the number of moves) is not known. The focus of attention is usually not so much on what is the best way to play such a game, but rather on whether one or the other player has a winning strategy.

An additional dichotomy is whether the players are in complete conflict, where the model employed is a non-cooperative one, or they have some commonality, where a more cooperative game model may be more appropriate. Such commonality could be, for instance, that the players by cooperating increase their individual payoffs or even that the players participate in groups and the payoff is given to a group and not to an individual player. Therefore, in a *cooperative game* the players are able to form binding commitments. In *non-cooperative games* this is not possible. Of the two types of games, non-cooperative games, and by non-cooperative we refer to the case where the players in the game are antagonistic and the individual payoffs are modelled in detail, are able to model situations to the finest details, producing more accurate results on the individual level, whereas cooperative games, situations where the group payoff is studied instead of the individual, focus on the game at large. Hybrid games contain cooperative and non-cooperative elements. For instance, coalitions of players are formed in a cooperative game, but these play in a non-cooperative way. It is important to note here that cooperative behavior can arise between antagonistic players in games with non-cooperative nature of play, and we will examine some of these situations in subsequent chapters.

Another important categorization is whether we are dealing with a game where the players have *complete information* about all actions taken or only *partial information*. Finally, there are the games of *perfect information* versus the ones of *imperfect information*. A game is one of perfect information if all players know the moves previously made by all other players. Thus, only sequential games can be games of perfect information, since in simultaneous games not every player knows the actions of the others. Perfect information is often confused with complete information, which is a similar concept. Complete information requires that every player knows the strategies and payoffs available to the other players but not necessarily the actions taken. Games of incomplete information can be reduced, however, to games of imperfect information by introducing *moves by nature* [9].

When characterizing a game it is important to keep in mind the various possible categorizations of game models in order to better describe the required strategic situation as completely as possible.

Game theory[1] appeared formally in the 1940s in a text by John von Neumann and Oskar Morgenstern [15], although the ideas of games and equilibria are found as early as 500 AD in the Babylonian Talmud, which is the compilation of ancient law and tradition for the Jewish religion [21], as well as in the 1800s in Darwin's *The Descent of Man, and Selection in Relation to Sex* [22]. Game Theory grew more popular in the 1950s and the 1960s with important contributors such as John F. Nash [16, 17, 18], Thomas Schelling [19], Robert John Aumann [20] and John Harsanyi [23, 24], as well as in the 1970s with Reinhard Selten [25], giving all the above scientists Nobel Prize awards in Economics in 1994 (John F. Nash, John Harsanyi and Reinhard Selten) and in 2005 Robert John Aumann and Thomas Schelling. It is worth mentioning some more recent, important contributors of Game Theory such as Ariel Rubinstein who contributed in the theory of bargaining [26] and Vincent P. Crawford for his work with redefinition of equilibria [27].

It is mentioned earlier that game models employ the element of rationality. However, one may argue that rationality implies that the players are perfect calculators and flawless followers of their best strategies, which is not always a correct replicate of a particular situation, thus this may not always be the case. Therefore, rationality may be better described to be the players' knowledge of their own interests based on each player's own value system. Based on this element of rationality the players calculate their possible strategies. Depending on whether the game is normal or extensive, strategies may consist of single actions or sequences of actions, and each strategy gives a complete plan of action, considering also reactions to actions that may be taken by the opponent. Strategies may be pure, i.e., provide complete definitions of how a player will play in the game (his moves), or mixed, i.e., assignments of a probability to each pure strategy.

Games are motivated by *profitable* outcomes that await the players once the actions are taken. These outcomes are referred to as payoffs. Payoffs for a particular player capture everything in the outcomes that the particular player cares about. If a player faces a random prospect of outcomes, then the number associated with this prospect is the average of the payoffs associated with each component outcome, weighted by their probabilities.

The solution to a strategic game is derived by establishing *equilibria*. Equilibria may be reached during the interaction of players' strategies when each player is using the strategy that is the best response to the strategies of the other players (i.e., given the strategies of the other players, the selected strategy results in the highest payoffs for each player participating in the game). The idea of equilibrium is a useful descriptive tool and furthermore, an effective organizing concept for analyzing a game theoretic model. For normal

---

[1]The text on Game Theory in this book is mainly based on [10, 11, 12, 13, 14].

form games the Nash Equilibrium is used as a solution concept, where every player's action is the best response to the actions of all the other players. For sequential-moves games, e.g., repeated games, the equilibrium used is known as the subgame perfect equilibrium or the rollback equilibrium (for finite repeated games). In such games the players must use a particular type of interactive thinking; players plan their current moves based on future consequences considering also opponents' moves. Therefore, the equilibrium in such a game must satisfy this kind of interactive thinking, and subgame perfect equilibrium does exactly that, by planning the best responses for every possible subgame or interaction.

This book considers the interactions in communication networks, with particular discussion on the interactions between the user and the access networks available to the user, or the interactions between the access networks themselves. Describing and analyzing entity interactions is a situation that makes a good candidate to be modeled using the theoretical framework of Game Theory. Game Theory provides appropriate models and tools to handle multiple, interacting entities attempting to make a decision and seeking a solution state that maximizes each entity's utility. Game Theory has been extensively used in networking research as a theoretical decision making framework, e.g., for routing [1, 2], congestion control [3, 4], resource sharing [5, 6], and heterogeneous networks [7, 8]. These interactions may benefit from such a theoretical framework that considers decision-making, interacting entities.

We advocate that the game theoretical framework is suitable for generating profitable behaviors/strategies for interacting entities in conflicting situations, and we explore its application upon seemingly conflicting interactions, as for example those occurring in converged heterogeneous communication networks. We structure the book so that each chapter describes and analyzes a typical example of such an interactive situation in a networking environment, using existing game theoretical models, and theoretical conclusions are drawn for each interactive situation; the theoretical conclusions are further reinforced with appropriate numerical results. Overall, these conclusions show that cooperation can indeed be motivated in the selected interactive situations and furthermore, that this cooperation is beneficial for the interacting entities. Cooperation appears to be a beneficial solution both in node-to-node interactive situations, as well as in network-to-network interactions (e.g., in Next Generation converged communication networks). We show that through the use of a game theoretic framework, cooperation can be motivated. Overall, Game Theory in general, and cooperation in particular have been used to solve various networking problems. In this book we turn our focus mainly on interactions in Next Generation Communication Networks; however, applicable cooperation scenarios can be expected in several other communication problems, including node-to-node interactive situations.

On a more practical note, in order to be able to enforce these solutions in a real heterogeneous communication network, additional issues must be considered such as the possible architecture that would enable easier management of the heterogeneity of the system, the repeatability of the interaction, which is the case in the considered scenarios, and the compensation set. The architecture considered to host such model in a mobile communication network is envisioned to be a multi-entity system, where a platform administrator (either a centralized or a distributed process) has knowledge of all participating access networks joined to a common core network, which is either IP-based with SIP signalling, or supporting a fully implemented IP Multimedia Subsystem (IMS) infrastructure to ensure multimedia support over all participating networks in an access-agnostic manner. Such architecture would support the existence of several autonomous entities (e.g., content and context providers, network operators, etc.) motivating the overall architecture to be more user-centric instead of network-centric, since all these entities have a common goal of satisfying the user in order to receive the appropriate compensation/payment as for instance, in the user-network interaction model presented in Chapter 2. An initial study of this appears in [29].

Furthermore, the existence of several autonomous entities acting independently requires the existence of certain policies which may integrate the interests of these entities by enforcing some rules for the better management and operation of the overall system. Such policies may deal with setting the compensation, i.e., the payment, corresponding to a particular quality level of a requested service. This has to do with the range of qualities in which a particular content (e.g., a video) is available and the corresponding costs and allowable profits. Policies may have to do with strategy and profile configurations *enforced* onto users and networks by the platform administrator process. To achieve the repeatability, which is a major element in these strategies, there arises the need for the existence of a variable as part of the internal logic of a user terminal or a network gateway node such that both the user and the network may *remember* the previous action of the opponent entity. Thus, ideas of reward and punishment, elaborated in the theoretical model, may be implemented.

1.1   Book structure

In structuring the book we adopt the following approach. For each chapter, we firstly introduce basic theory for dealing with a particular interactive situation and then, in the second part of each chapter, present example scenarios, showing the applicability and power of the theory. This book does not attempt to provide a comprehensive toolbox of game theory, rather it attempts to show how particular aspects of game theory can be usefully employed to formulate and solve a number of representative interactive situations commonly appearing in the field of communication networks. In the second part of each chapter, we adopt the scenario's view to demonstrate a number of cooperative interactions and discuss how these could be addressed within the theoretical framework presented in the first part of the chapter. Next, we summarize the book contents.

We begin with the interaction between a user and a network, where the cooperative nature is motivated by an infinitely repetitive game, and appropriate strategies for both the user and the network are evaluated in order to select the ones that achieve strong motivation for the two entities toward cooperating and remaining in cooperation; an adaptive user strategy and the corresponding game profile are also illustrated in Chapter 2. Next, the bargaining situation between two access networks attempting to partition a service payment optimally, is modelled and resolved using the Nash Bargaining Game model, and accordingly the Nash Bargaining Solution, which is equivalent to the immediate resolution of the widely used Rubinstein Bargaining Game. Truthfulness is an issue that must always be considered in such bargaining situations, and thus a Bayesian game model leads us toward inducing truthfulness from the participating players in Chapter 3. In Chapter 4 the notion of group strategies is introduced to motivate cooperation in a situation where many topologically proximal players coexist and are faced with the dilemma of entering group cooperation or defecting from cooperation, and we show that there exists motivation to enter and remain in a cooperative group and employ the group strategy increasing both the group and the individual payoffs. Finally, regarding the coalition formation process in the case of increased service demand, we explore the conclusions of the analytic study stemming from the assumptions presented in Chapter 5, and move on to present an appropriate payoff scheme through what we refer to as the *Popularity Power Index*. The payoffs that result from the employment reward the coalitions, which are more likely to be formed in accordance with their popularity. Therefore, the game resolution favors a fairer payoff allocation between the participating access networks.

The following paragraphs discuss in more detail the interactive situations presented in Chapters 2, 3, 4 and 5 and we also present Chapter 6, the implementation chapter. In fact, the technical chapters of the book, Chapters 2, 3, 4 and 5, are organized as follows: in each chapter we present the basics of the game theoretic tools that will be used to model and analyze a particular scenario and then follow up with a representative example and its analysis. The first two chapters deal with two-player interaction while the next two chapters deal with interactions between multiple players.

The first selected interactive situation presented as the scenario of Chapter 2 illustrates various cooperative aspects of the relationship between the user and the network when they interact in the following scenario: *the user participates in a User Generated Content service supported by the network, where the user becomes the independent content provider.* Through such services, the user acting as a content provider uses the network infrastructure to distribute audio and video content to the community of subscribers. A network operator supporting User Generated Content services, attracts users by offering users a quick and easy way to socialize user generated audio and video content. Hence, given that both the user and the network have incentives to participate in User Generated Content services, cooperation between the user and the network is desirable. The interaction has been modelled using game theoretic tools, in an attempt to model an interaction that motivates cooperation. Firstly, the interaction is modelled as a one-shot game, and it is shown that there exists equivalence of a one-shot game model of user-network interaction to the one-shot *Prisoner's Dilemma Game*. This is an important finding, since the *Prisoner's Dilemma* is known to lead to a cooperative solution under certain conditions, as for example a repeated game, motivating this chapter to explore this interaction further toward the direction of cooperation. In this chapter we demonstrate how motivation of cooperation between the two entities is achieved through a *repeated game* model of the user-network interaction. Exploring existing and new strategies for the two entities, we see that when the strategies used by the players of the repeated user-network interaction model involve punishment to motivate cooperation, then harsher punishments motivate cooperation more easily. However, since practically a user wouldn't choose to employ the harshest punishment, i.e., leaving the interaction forever, we develop a strategy that uses an *adaptive* punishment method in the repeated user-network interaction game. The adaptive strategy motivates cooperation and achieves satisfying results in terms of motivation and in terms of payoffs, becoming the strategy of choice for a user when compared to the other strategies examined for the scenario. As a consequence, a profile of the repeated user-network interaction game where the user employs the adaptive punishment method and the network employs the well-known *Tit-for-Tat* strategy, generates the most profitable payoffs for both players.

The selected interaction scenario presented in Chapter 3 involves two networks interacting to support a service with additional quality guarantees (e.g., for a premium user). Therefore, the first network supports the service and the second network reserves the appropriate resources to ensure service continuity, in the case that the first network demonstrates quality degradation so that the user is guaranteed a seamless experience, e.g., in terms of service continuity at similar QoS levels. The two networks must cooperate to partition the service payment, since the fact that two networks support the service is transparent to the user offering the service payment. This interaction between two networks has been modelled using game theoretic tools, in such a way as to motivate cooperation in terms of partitioning the available payment for the particular service. The proposed payment-partition model is shown to be equivalent to the well-known *Rubinstein Bargaining Game*, if the agreement in the Rubinstein Bargaining Game is reached from the first negotiation period. In addition, the chapter shows that there exists equivalence between the payment-partition game and the *Nash Bargaining Game*, due to the equivalence of the Nash bargaining game to the Rubinstein Bargaining Game, if the agreement in the Rubinstein Bargaining Game is reached from the first negotiation period. Thus, an optimal solution for the payment-partition game exists and is based on the *Nash Bargaining Solution*, resulting in a partition determined by the cost each of the two networks has for supporting the service.

As a side result, it has been shown that if a constant probability of demonstrating quality degradation is included in the networks' payoff functions for the payment-partition game, this does not affect the optimal partition proposed by the Nash Bargaining Solution. However, it affects the payoff of the individual networks. This is important since the technology employed by each network and the established quality provisioning mechanisms are always susceptible to QoS degradation due to the dynamic nature of the heterogeneous network and the mobility of the users; considering more realistic conditions for the network models, i.e., a constant degradation probability (according to the technology and quality provisioning mechanisms for each network), adds credibility to the solution. Once the optimal partition is determined, the payment-partition game is modelled as a one-shot *Bayesian* game, to investigate truthfulness on behalf of the participating networks regarding their own costs, since the declaration of costs is very important in determining the optimal partition. It has been shown that no matter whether a network believes that the opponent network has declared lower or higher cost than its own cost, it is still motivated to lie about its real costs. To motivate truthfulness, a *pricing mechanism* can be used in the payoffs of the users, which we show to work very effectively toward motivating the networks to be truthful.

In Chapter 1 the networking scenario considers a dense urban residential area where each house/unit has its own wireless access point (AP), deployed without any coordination with other such units. Lacking any control regarding the efficient utilization of the communication channel, it is quite common for a terminal served by one of the APs to be within the signal range of multiple alternative APs. Since all APs are in competition for the same communication resource (radio channel), and the current standards dictate that at any given time every terminal must be rigidly associated with one particular AP, this situation results in increased interference and consequently a low utilization efficiency of the radio resource. In a dense deployment, it would be much better for individual APs that are in physical proximity to each other to form groups, where one member of the group would serve the terminals of all group members in addition to its own terminals, so that the other access points of the group can be silent or even turned off, thereby reducing interference and increasing overall Quality of Experience (QoE). Since there is no centralized entity that can control the APs and force them to form cooperative groups, the creation of such groups must be able to arise from a distributed process where each AP makes its own decisions independently and rationally for the benefit of itself and its terminals. The scenario models the idea of cooperative neighborhoods as a game and shows that a group cooperative strategy in equilibrium, i.e., a strategy for units to voluntarily participate in a group where members serve terminals on a rotating basis, has the property that a unit participating in the group strategy is more likely to gain more in terms of QoE, than a unit defecting from such cooperation. In fact, we illustrate how a protocol, with point of operation the game theoretic equilibria of the game, can maintain robustness against uncoordinated deployments in dense residential areas. We refer to this as the *cooperative-neighborhood game.*

In Chapter 5, the selected scenario involves the interaction between multiple networks when they must cooperate to serve a large service demand that is best served by more than one network (for example, if none of those networks can serve the demand on its own). The coalition formation process between multiple networks is modelled as a game we refer to as the *Network Synthesis* game, in which individual networks with insufficient resources form coalitions in order to satisfy service demands. The Network Synthesis game is equivalent to the well-known *Weighted Voting Game.* This equivalence encourages the use of power indices for payoff allocation, as often used to resolve the Weighted Voting Game. A comparative study of well-known *power indices* representing payoff schemes, is provided for the network synthesis game. Based on conclusions from analyzing these schemes, the chapter presents a new power index, the *Popularity Power Index* (PPI), which associates the popularity of each network to the number of stable coalitions it participates in. This power index achieves fairness, in the sense that it only considers the possible coalitions that would be formed if payoffs were assigned proportionally to the networks' contributions. An analysis of the coalition formation is provided for both transferable and non-transferable payoffs, in order to determine stable coalitions using the *core* and *inner core* concepts. The most appropriate power index for the network synthesis game, from the existing power indices investigated, is a power index that provides stability under the core concept, known as the *Holler-Packel Index (HPI).* The core and inner core equilibrium concepts are further investigated to show that coalitions that would be formed using the newly proposed PPI to assign payoffs, are only coalitions that would be stable under the inner core concept. Therefore, the PPI provides a simple and fair payoff allocation method that is equivalent to a stable cooperative equilibrium solution of the Network Synthesis Game.

In Chapter 6 we present implementation of the Iterated Prisoner's Dilemma in MATLAB based on an existing implementation [3]. Using the existing framework we show the implementation of strategies used in Chapter 2 and in Chapter 4. We explain the programming details in an attempt to simplify the use and modification of the code for the interested reader. A pictorial view of how the book contents are arranged is presented in Figure 1.2 for easy reference.

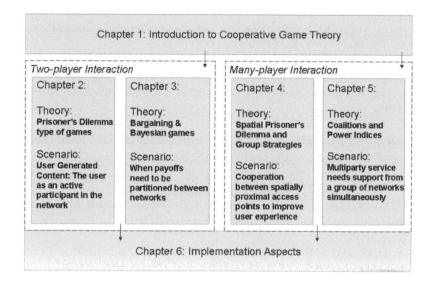

Figure 1.2: Book contents.

References

[1] A. van de Nouweland, P. Borm, W. van Golstein Brouwers, R. Groot Briunderink and S. Tijs, *A Game Theoretic Approach to Problems in Telecommunication*, Management Science, vol. 42, no. 2, pp. 294-303, February 1996.

[2] A. Orda, R. Rom and N. Shimkin, *Competitive Routing in Multiuser Communication Networks*, IEEE/ACM Transactions on Networking, vol. 1, no. 5, pp. 510-521, 1993.

[3] A. de Palma, *A Game Theoretic Approach to the Analysis of Simple Congested Networks*, The American Economic Review, vol. 82, no. 2, pp. 185-199, 2005.

[4] L. Lopez, A. Fernandez and V. Cholvi, *A Game Theoretic Comparison of TCP and Digital Fountain based protocols*, Computer Networks, vol. 51, pp. 3413–3426, 2007.

[5] S. Rakshit and R. K. Guha, *Fair Bandwidth Sharing in Distributed Systems: A Game Theoretic Approach*, IEEE Transactions on Computers, vol. 54, no. 11, pp. 1384–1393, November 2005.

[6] H. Yaiche, R. R. Mazumdar and C. Rosenberg, *A game theoretic framework for bandwidth allocation and pricing in broadband networks*, IEEE/ACM Transactions on Networking, vol. 8, no. 5, pp. 667–678,

2000.

[7] J. Antoniou, I. Koukoutsidis, E. Jaho, A. Pitsillides, and I. Stavrakakis, *Access Network Synthesis in Next Generation Networks*, Elsevier Computer Networks Journal Elsevier Computer Networks Journal, vol. 53, no. 15, pp. 2716-2726, October 2009.

[8] J. Antoniou, V. Papadopoulou, V. Vassiliou, and A. Pitsillides, *Cooperative User-Network Interactions in next generation communication networks*, Computer Networks, vol. 54, no. 13, pp. 2239-2255, September 2010.

[9] Leyton-Brown and Shoham, *Essentials of game theory: A concise multidisciplinary introduction*, Synthesis lectures on artificial intelligence and machine learning, vol.2, no.1 p. 1-88, 2008.

[10] R. B. Myerson, *Game Theory: Analysis of Conflict*, Harvard University Press, Cambridge, Massachusetts, 2004.

[11] H. Gintis, *Game Theory Evolving: A Problem-Centered Introduction to Modeling Strategic Interaction*, Princeton University Press, Princeton, New Jersey, 2000.

[12] A. Dixit and S. Skeath, *Games of Strategy*, W.W.Norton & Company, New York, 1999.

[13] A. Muthoo, *Bargaining Theory with Applications*, Cambridge University Press, Cambridge, UK, 2002.

[14] M. J. Osborne and A. Rubinstein, *A Course in Game Theory*, Massachussetts Institute of Technology, Massachussetts, USA, 1994.

[15] J. von Neumann and O. Morgenstern, *Theory of Games and Economic Bahavior*, Princeton University Press, Princeton, 1944.

[16] J. F. Nash, *The Bargaining Problem*, Econometrica, v. 18, no. 2, April, 1950, pp. 155–162.

[17] J. F. Nash, *Non-cooperative games*, Annals of Mathematics, v. 54, no.2, September, 1951, pp. 286–295.

[18] J. F. Nash, *Two-person cooperative games*, Econometrica, v. 21, no. 1, January 1953, pp. 128–140.

[19] T. C. Schelling, *The Strategy of Conflict*, Harvard University Press, Cambridge, Massachusetts, 1960.

[20] R. J. Aumann, *Acceptable Points in Games of Perfect Information*, Pacific Journal of Mathematics, v. 10, 1960, pp. 381–417.

[21] R. J. Aumann, *Game Theory in the Talmud*, Jewish Law and Economics Research Bulletin Series, 2003.

[22] C. Darwin, *The Descent of Man and Selection in Relation to Sex*, John Murray Publishing, vol. 1, London, 1871.

[23] J. C. Harsanyi, *Measurement of Social Power, Opportunity Costs, and the Theory of Two-Person Bargaining Games*, Behavioral Science, v. 7, 1962, pp. 67–81.

[24] J. C. Harsanyi, *Games with Incomplete Information Played By Bayesian Players, Parts I, II, and III*, Behavioral Science, v. 14, 1967, pp. 159–182, 320–334, 486–502.

[25] R. Selten, *Reexamination of the Perfectness Concept for Equilibrium Points in Extensive Games*, International Journal of Game Theory, v. 4, 1975, pp. 25–55.

[26] A. Rubinstein, *Perfect Equilibrium in a Bargaining Model*, Econometrica, v. 98, no. 1, 1982, pp. 97–109.

[27] V. P. Crawford, *Equilibrium without Independence*, Journal of Economic Theory, v. 50, no. 1, February, 1990, pp. 127–154.

[28] G. Taylor, *Iterated Prisoner's Dilemma in MATLAB*, Archive for the "Game Theory" Category, http://maths.straylight.co.uk/archives/category/game-theory, March, 2007.

[29] J. Antoniou and M. Stylianou and A. Pitsillides, RAN Selection for Converged Networks supporting IMS Signalling and MBMS Service, Proceedings of the 18th International Symposium on Personal, Indoor and Mobile Radio Communications (PIMRC '07), September 2007.

Chapter 2

Cooperation for two: Prisoner's Dilemma type of games

2.1 Introduction

This chapter investigates how to model and study the cooperative aspects between any two entities whose interaction satisfies the requirements of the well-known game theoretic interaction model of the Prisoner's Dilemma. The Prisoner's Dilemma is an important model because it models the interaction between two seemingly antagonistic entities *luring* them into cooperative behavior in the repetitive version of the game, even though in the one-shot game this does not appear as the most profitable choice for the involved players. However, even in the one-shot game, mutual cooperation is the most socially efficient choice, since it pays more than mutual defection, which, as we will later show, is the Nash Equilibrium for the one-shot game.

The *Prisoner's Dilemma* and *Iterated Prisoner's Dilemma*[1] have been a rich source of research material since the 1950s. However, the publication of Axelrod's book in 1984 [2] was the main driver that brought this research to the attention of other areas outside of game theory, as a model for promoting cooperation.

Prior to emerging into the details of the Prisoner's Dilemma Game and applying it to a networking scenario, we offer an overview of similar two-player games that model interactions between two antagonistic entities that have the same two options: to cooperate or to defect. We hence justify the Prisoner's Dilemma game as our chosen game for a situation that aims to promote cooperation between the two antagonistic players.

---

[1] The text on Prisoner's Dilemma is based on [5].

2.2   Prisoner's Dilemma and similar two-player games

The Prisoner's Dilemma is basically a model of a game, where two players must decide whether to cooperate with their opponent or whether to defect from cooperation. Both players make a decision without knowing the decision of their opponent, and only after the individual decisions are made, these are revealed. The story behind this model is the following: *Two suspects are arrested by the police. Since the policemen have insufficient evidence for their conviction, they separate the two suspects and offer them a deal. If one testifies against the other (i.e., defects from the other) and the other remains silent (i.e., cooperates with the other), the betrayer goes free and the silent goes to jail for 10 years. If both remain silent, then they both go to jail for only 6 months with a minor charge. If both testify against each other, each gets a 5-year sentence.* Each suspect must choose whether to testify or to remain silent, given that neither learns about the decision of the other until the end of the investigation. What should they do?

Mutual cooperation has a reward for both of receiving the punishment which is the least harsh. However, such decision entails the risk that in case the other player defects, then the cooperative player will receive the harshest punishment of 10 years. Given the risk of cooperation, it is very tempting to defect because if the opponent cooperates, then defecting will result in the best payoff, which is to go free, although, if the other opponent also defects then an average punishment will be received by both (i.e., 5 years). The decision of what to do comes from the following reasoning: *If a player believes that his opponent will cooperate, then the best option is certainly to defect since the payoff will be to go free. If a player believes that his opponent will defect, then by cooperating he takes the risk of receiving the harshest punishment, i.e., 10 years, thus the best option is again to defect and share the average punishment of 5 years with his opponent.* Therefore, based on this reasoning, each player will defect because it is the best option no matter what the opponent chooses. However, this is not the best possible outcome of the game. The best solution would be for both players to cooperate and receive the least harshest punishment of 6 months.

The desirable cooperative behavior must be somehow motivated so that the players' selfish but rational reasoning results in the cooperative decision. In fact, what we have described is a *one-shot* Prisoner's Dilemma, i.e., the players have to decide only once – no previous or future interaction of the two players affects this decision. The payoffs to the two players can be summarized in the following way illustrated in Table 2.1: Payoff A represents defection when the opponent (Player 2) cooperates, payoff B represents mutual cooperation, payoff C represents mutual defection and finally payoff D represents cooperation when the opponent defects. According to the storyline of the game, the relationship between the payoffs is such that $A > B > C > D$. Another constraint in the payoffs that must be kept in the unfolding of a Prisoner's Dilemma game is that the payoff for mutual cooperation must be greater than the payoff of alternatively choosing to cooperate when the opponent defects and defect when the opponent cooperates, i.e., $\frac{A+D}{2} < B$. Any numerical values that obey the two payoff relationships are valid payoffs for any Prisoner's Dilemma type of game.

Table 2.1: General payoffs for the Prisoner's Dilemma

	Player 2 Cooperates	Player 2 Defects
Player 1 Cooperates	B_1, B_2	D_1, A_2
Player 1 Defects	A_1, D_2	C_1, C_2

Let us consider the numerical values offered in the above storyline. Table 2.2 demonstrates these values. To find a Nash equilibrium, i.e., a set of

Table 2.2: Example payoffs for the Prisoner's Dilemma game

	Player 2 Cooperates	Player 2 Defects
Player 1 Cooperates	$-0.5, -0.5$	$-10, 0$
Player 1 Defects	$0, -10$	$-5, -5$

actions that when played none of the players has an incentive to change, we look at the rows and columns and select the cell or cells in our table where each player, given any of the opponent's possible actions, has the greatest possible payoff. The unique Nash equilibrium in this game is the cell where both Player 1 and Player 2 defect and each receive a payoff of -5. In fact, defection is a dominant strategy choice for each player, i.e., no matter what the opponent chooses to play, defection always results in a highest payoff. However, we notice that mutual defection is not the most profitable course of action in the end since if both players cooperate, mutual cooperation will result in a payoff of 0.5, which is much preferrable to -5. Thus, we say that mutual cooperation is the most socially efficient course of action in this game, even though the Nash equilibrium points to mutual defection.

In the subsequent paragraphs we describe games that similarly to the Prisoner's Dilemma use payoffs A, B, C and D as those are shown in Table 2.1, but the relationship of these payoffs are different in each game. The relationship is explained separately for each game.

The next game we will briefly look at, which deals with the antagonistic interaction of two players, is the Dead Lock game. In this game, similarly to the Prisoner's Dilemma game, the two players have the options to cooperate or to defect. The relationship between the payoffs for these options obeys the constraint $A > C > B > D$, where as before A represents defection when the opponent cooperates, B represents mutual cooperation, C represents mutual defection and D represents cooperation when the opponent defects. Now, the payoff for mutual defection exceeds the payoff for mutual cooperation. In this game the second payoff constraint is not imposed.

Table 2.3 gives a numerical example for this game: Again, we can see that

Table 2.3: Example payoffs for the Dead Lock game

	Player 2 Cooperates	Player 2 Defects
Player 1 Cooperates	1, 1	0, 3
Player 1 Defects	3, 0	2, 2

the Nash Equilibrium for the game is when the two players select the mutual defection course of action and furthermore, we observe that defection is the dominant strategy for both players. However, unlike the Prisoner's Dilemma game, in the Dead Lock game, mutual defection is actually better in terms of payoffs than the mutual cooperation and thus not only is it the Nash Equilibrium for the game but it is also the most socially efficient outcome for this game.

We now move on to describe what is known as the Chicken game. The name of the game comes from the following story: Two drivers race toward each other in a game of chicken, as it is called, where they drive until one of them chickens out and swerves. In this case cooperation is to swerve, whereas defection is to drive straight. If a player drives straight and the other swerves, the first wins the highest payoff, while the second loses the race. If they both swerve, mutual cooperation is the second largest payoff, neither of them wins the race and neither of them loses. The worst outcome for both is to both defect and drive straight, in which case this will result in a head on collision, the lesser payoff for both. Therefore, the relationship between payoffs A, B, C and D, as these are described above is such that $A > B > D > C$. We illustrate indicative numerical values for the Chicken game in Table 2.4

Table 2.4: Example payoffs for the Chicken game

	Player 2 Cooperates	Player 2 Defects
Player 1 Cooperates	2, 2	1, 3
Player 1 Defects	3, 1	0, 0

There are two Nash equilibria in this game, if player 1 defects and player 2 cooperates or if player 1 cooperates and player 2 defects. However, there is no dominant strategy choice for either of the players. The mutual cooperation is a socially efficient option in this game, but so are the two equilibria options, therefore the players are not motivated enough to cooperate.

The last two-player game that we will be looking at in this section is called the Stag Hunt game, and it is also known as the trust dilemma or the assurance game. The stag is an adult deer and the game is inspired by the tradition of deer hunting. This game describes the conflict between safety and social cooperation and it was first inspired by a French philosopher, Jean Jacques Rousseau. Since hunting deer is a challenging task it requires two hunters to cooperate and stay in position for the stag hunt. What happens when a running rabbit passes by? Do the hunters stay in position, i.e., cooperate, or follow the temptation of hunting the rabbit, i.e., defect from cooperation? In this case, mutual cooperation will bring the highest payoff to both players, while the worst case scenario is for one player to defect and the other to stay in position, because in this case he cannot be successful on his own. The defector whose opponent cooperates gains a payoff not as high as if he would remain in cooperation and not as low as the payoff if both hunters defect. Therefore, the relationship between payoffs A, B, C, and D is such that $B > A > C > D$ and are illustrated in terms of numerical values in Table 2.5. In this game both

Table 2.5: Example payoffs for the Stag Hunt game

	Player 2 Cooperates	Player 2 Defects
Player 1 Cooperates	3, 3	0, 2
Player 1 Defects	2, 0	1, 1

mutual cooperation and mutual defection are Nash Equilibria. Neither player has a dominant strategy choice, in fact each player is better off to mimic his opponent's actions, i.e., cooperate if the opponent cooperates and defect if the opponent defects. The optimal choice is mutual cooperation, which is also the socially efficient course of action by the players.

Remark 1. *We observe that the relationships between the payoffs plays an important role in the evolution of a two-player game where the two players have the same two options: cooperate and defect. Each one of these games demonstrates a different situation where the cooperation and defection actions are valued differently and thus affect the players into different decisions. In using any of these games to model a given scenario, one must consider the equivalency of the scenario payoffs to the game payoffs in terms of the payoff relationships, as well as the goal of the scenario to promote cooperation. It is obvious that only the first scenario motivates cooperation in terms of the only socially efficient option, while the other scenarios do not support uniquely the option of mutual cooperation. The scenario that we demonstrate in this chapter aims to promote cooperation. Therefore we select the Prisoner's Dilemma game and show the equivalence of the payoff relationships to our proposed scenario, and that the cooperation it promotes is a socially efficient choice. Furthermore, to promote the most profitable choice for each of the players given any of the opponent's actions (i.e., be the Nash Equilibrium of the game), we use the infinitely repeated version of the Prisoner's Dilemma game, also known as the Iterated Prisoner's Dilemma. Details on the game model itself and on how to use it to demonstrate that mutual cooperation is an efficient solution for the proposed scenario are given in the rest of the chapter.*

2.3   Focusing on Prisoner's Dilemma

Definition 2.3.1 summarizes the actions and constraints of any two players
interacting for a single round (i.e., *one-shot* game) of the Prisoner's Dilemma
type of game.

Definition 2.3.1 (Prisoner's Dilemma type of game). *[5] Consider an one-
shot strategic game with two players in which each player has two possible
actions: to cooperate with his opponent or to defect from cooperation. Furthermore, assume that the two following additional restrictions on the payoffs
are satisfied:*

1. *The order of the payoffs is shown in Table 2.1 for each player $j \in \{1,2\}$
   and is such that $A_j > B_j > C_j > D_j$.*

2. *The reward for mutual cooperation should be such that each player is not
   motivated to exploit his opponent or be exploited with the same probability, i.e., for each player it must hold that $B_j > \frac{A_j+D_j}{2}$.*

Then, the game is said to be equivalent to a Prisoner's Dilemma type of game.

Although the one-shot game motivates the players to defect from cooperation, cooperation may still evolve from playing the game repeatedly, against
the same opponent. This is referred to as Iterated Prisoner's Dilemma, which
is based on a repeated game model with an unknown or infinite number of
repetitions. The decisions at such games, which are taken at each repetition of
the game, are affected by past actions and future expectations and can result
in strategies that motivate cooperation.

The model of a repeated game is designed to examine the logic of long-term
interaction. It captures the idea that a player will take into account the effect
of his current behavior on the other player's future behavior. Repeated game
models aim to explain phenomena like cooperation, threats and punishment.

The repeated game models offer insight into the structure of behavior
when individuals interact repeatedly, a structure that may be interpreted in
terms of social norm. The results show that the social norm needed to sustain
mutually desirable outcomes involves each player threatening to *punish* any
other player whose behavior is undesirable, outside the social norm. Each
player uses *threats* to warn the opponent that such punishment may follow.
The opponent's actions will be influenced by whether the threats are credible
and if there is sufficient incentive for the player to carry out his threat. Thus,
punishment depends on how players value their future payoffs. Furthermore,
the punishment can be as harsh as lasting forever, or as mild as lasting for
only one iteration.

A repeated game model can be one of two kinds: the horizon may be
finite, i.e., it is known in how many periods the game ends, or *infinite*, i.e.,
the number of game periods is unknown. The outcomes in the two kinds of
games are different. For instance, analyzing a finite version of the Prisoner's

Dilemma ends in the conclusion that the players are motivated to cheat as in the one-shot Prisoner's Dilemma, whereas an infinite version of the Prisoner's Dilemma results in a motivation for both players to cooperate. Since we are dealing with repetitive interaction, we will define a player's *strategy* to be a sequence of actions, one for each iteration of the game, until the game ends. The selection of a strategy for each of the two interacting players, defines the interaction of the two players for each round of the game, and hence for the total number of iterations that the game consists of. The combination of any pair of strategies, one for each player, is referred to as a *profile* for the game.

The Iterated Prisoner's Dilemma is quite a popular repeated game model. It demonstrates how cooperation can be motivated by repetition (in the case the number of periods is unknown), whereas in the one-shot Prisoner's Dilemma as well as in the finite version of the Iterated Prisoner's Dilemma, the two players are motivated to cheat. The main idea is that if the game is played repeatedly, then the mutually desirable cooperative outcome is stable because any deviation will end the cooperation, resulting in a subsequent loss for the deviating players that outweighs the payoff from the finite horizon game (horizon of one or more periods). Thus, when applying the model of a repeated game to a specific situation or problem, we must first determine whether a finite or infinite horizon is appropriate, based on the characteristics of the realistic situation we are modelling.

2.3.1 Motivating cooperation from repetition

A repeated game makes it possible for the players to condition their moves on previous history of the various stages, by employing strategies that define appropriate actions for each period. Such strategies are called *trigger strategies* [7, Chapter 6]. A trigger strategy is a strategy that changes in the presence of a predefined trigger; it dictates that a player must follow one strategy until a certain condition is met and then follow a different strategy, for the rest of the game.

One of the most popular trigger strategies is the *Grim Trigger strategy* [8, Chapter 11], which dictates that the player participates in the relationship in a cooperative manner, but if *dissatisfied* for some known reason, leaves the relationship forever. Given such a strategy, players have a stronger incentive to cooperate, since they face the threat of terminating the interaction. Such threats secure compliance of the opponent toward cooperation. Another popular strategy used to elicit cooperative performance from an opponent, is for a player to mimic the actions of his opponent, giving him the incentive to play cooperatively, since in this way he will be rewarded with a similar mirroring behavior. This strategy is referred to as *Tit-for-Tat* strategy [8, Chapter 11]. The subsequent scenario (Section 2.4) employs the Grim strategy and the Tit-for-Tat strategy as possible cooperative strategies, since the strengths of each strategy mentioned above are considered appropriate for the interaction presented in our scenario. In addition a defecting strategy is defined for each player separately, so that we can investigate corresponding game profiles, e.g., when both players use cooperative strategies or when the combination of strategies (the game profile) includes defecting strategies. The scenario is used as an example to show the model and calculations that lead to cooperation as the best strategy for both players. The idea of the *present value* (Section 2.3.2) is key to these calculations.

2.3.2   Present value

Examining the two-players relationship, we consider a signal that runs through the repeated game and may stop the interaction (such a signal could represent a *non-strategic* event that would force either of the players to terminate the interaction). Therefore, there is always a slight probability that the game will not continue in the next period. Let this probability be denoted as p.

In order to compare different sequences of payoffs in repeated games, we utilize the idea of the *present value of a payoff sequence* [8, Chapter 11], and we refer to it as the *Present Value* (PV). PV is the sum that a player is willing to accept currently instead of waiting for the future payoff, i.e., accept a smaller payoff today that will be worth more in the future, similar to making an investment in the current period that will be increased by a rate r in the next period.

Therefore, if a player's payoff in the next period were equal to 1, today the payoff a player would be willing to accept would be equal to $\frac{1}{1+r}$. If there is a probability that the game will not continue in the next period, equal to p, then the payoff a player is willing to accept today, i.e., the player's PV, would be equal to $\frac{1-p}{1+r}$. Let $\delta = \frac{1-p}{1+r}$, where $\delta \in [0,1]$ and often referred to as the *discount factor* in repeated games [8]. Given a payoff X in the next period, its PV in the current period equals $\delta \cdot X$.

Now, for an infinitely repeated game, a PV should include the discounted payoff of all subsequent periods of the game. Let the payoff from the current period be equal to 1. Then, the additional payoff a player is willing to accept for the next period equals to δ, for the period after that the additional payoff equals to δ^2 and so on. Thus, PV equals to $1 + \delta + \delta^2 + \delta^3 + \delta^4 + \ldots$, which, according to the sum of infinite geometric series, equals to $\frac{1}{1-\delta}$. Therefore, for a payoff X payable at the end of each period, the present value in an infinitely repeated game equals to $\frac{X}{1-\delta}$.

In order to determine whether cooperation is a better strategy in the repeated game for both users, we utilize the PV and examine for which values of $\delta = \frac{1-p}{1+r}$ a given strategy is a player's best response to the other player's strategy. A strategy in an infinitely repeated game gives the action to take at each decision node. At each decision node, i.e., for each period a player has the choice of two actions: either to take the risk and cooperate with the opponent, or to defect from cooperation.

2.4 Threats and punishments: The user as an active participant in the network

This section offers an example where a user-network interaction model is modelled as an Iterated Prisoner's Dilemma. It is shown that when threats of punishment are a part of each player's strategy, cooperation is motivated in the infinitely repeated interaction. Various strategies are given as examples where we observe the willingness of the players to keep cooperating. It is worth pointing out that there is greater motivation to cooperate when the punishment threat is bigger. However, normally it is more practical to avoid very big or very small punishments. Therefore, we propose the use of an adaptive state which reacts according to the player's behavior, rewarding players that are mostly cooperative, whereas punishing the ones that are not, progressively modifying the punishment according to past behavior.

Before presenting the theoretical formulation of such a class of interactive games, we present an illustrative example where such an interactive game can take place and then illustrate using analysis various forms of strategies, analysing each one with regards to the cooperation induced.

2.4.1 An illustrative scenario

The user-centric paradigm employed in Next Generation Networks (NGNs) has brought about the decoupling of content and carriage, with the mobile network operators taking the role of the content carrier, whilst the task of content provisioning is being handled by independent content providers. This decoupling is directly in accordance with one of the fundamental aspects of NGNs [1], the *decoupling of service provision from network and provision of open interfaces.*

Mobile networks participating in Next Generation mobile network environmnents are called upon to handle, on the one hand, the grand speed at which technology develops and, on the other hand, the ever-changing content pools. Thus, new interactive services invite users to participate actively in the network, changing the mobile subscriber role from a passive customer into an active participant. User Generated Content (UGC) services [3], which depend on user contribution, are becoming more and more popular in the fixed Internet (Facebook, Flickr, YouTube), and are expected to dominate the mobile market as well. Through such services, the user acting as a content provider uses the network infrastructure to distribute audio and video content to the community of subscribers. A network operator supporting UGC services, attracts users by allowing them to become producers of their own content and distribute this content to other subscribers in a relatively inexpensive way. Thus, network operators are in a position to benefit from the increasing interest in UGC services, by attracting more customers, who are offered a quick and easy way to socialize user generated audio and video content. Hence, given that both the user and the network have incentives to participate in UGC services, cooperation between the user and the network is desirable.

In this section, we illustrate various cooperative aspects of the relationship between the user and the network when they interact in the following scenario: *the user participates in a UGC service supported by the network, where the user becomes the independent content provider.*

UGC services invite users to participate actively in the network, changing the mobile subscriber role from a passive customer into an active participant. Consequently, this model creates a new user-network relationship from which both entities have something to give and something to receive, i.e., an exchange relationship. The user gives a compensation to participate in the UGC service hosted by the network and receives the *satisfaction* from participating, and the network receives the compensation and gives the resources to enable this service, distributing the user generated content.

However, allowing the user to become a content provider, results in the network having no control over the *quality* of the uploaded content, with the issue becoming more serious in the case of infringing content[2]. The user could upload infringing content, which may cost in terms of reputation and fines (from possible copyright issues) to the network. Thus, the network is motivated to prevent such situation. For example, recognizing infringing content could be achieved through the setup and maintainance of a secure framework aiming to thwart infringement by employing content identification technology [4], in order to match newly uploaded content against existing copyrighted content. A broker entity could undertake this task and provide networks the option of checking the *originality* of contents uploaded by their users, through the use of this content identification framework, for a certain fee per request. Such techniques are explicitly mentioned in a list of principles for user generated content posted on the web [14] and supported by the world's leading Internet and media companies, including CBS Corp., Dailymotion, Fox Entertainment Group, Microsoft Corp., MySpace, NBC Universal, Veoh Networks Inc., Viacom Inc. and The Walt Disney Company.

In particular, we identify the interaction between the user and the network as the following:

The user wants to participate in the UGC service and upload some own audio or video content. To participate in the service, the user must make a UGC service request to the network. For simply participating in the service, the request is referred to as a *basic* service request. In addition to this, the user may request *content identification* of the uploaded audio or video content, i.e., request that the content is checked and identified as *non-infringing*, content. This may be referred to as an *enhanced* service request. Note that for the *basic* service case, the user is expected to *assure* the network that the content is not copyright-infringing. In this case the user may or may not be truthful.

---

[2]One could also include abusive or offensive content in this scenario. Similar argumentation applies.

The payment of the user for participating in the *basic* service is less than the payment to participate in the *enhanced* service, because content identification entails additional cost for the network, which needs to request such identification from a *copyright broker* entity[3]. We assume that, for content identification, the network entails a certain cost per request by the copyright broker.

Let the user's payment to the network be referred as compensation, which is either given for a *basic* UGC service request or for an *enhanced* UGC service request. The user has an incentive to request the *basic* service in order to save money, but simultaneously he takes the risk that if the content is checked and identified as infringing, he will be held liable (with probably higher costs in terms of fines). On the other hand, if the user requests an *enhanced* service, he will pay more but at the same time he minimizes the risk that he will be held liable if the content is identified as infringing.

The decision dilemma on the user side is not the only one in this interaction. There exists a decision dilemma on the network side as well. The network has to decide whether or not to check the uploaded content, knowing that checking entails an additional cost for the network, but at the same time minimizes the risk for liability of the network in case the content is identified as infringing. The network has an incentive not to check the content in order to avoid the additional cost, but at the same time it risks being held liable for infringing content, which may in turn lead to even greater costs in terms of suits and fines.

Next, we summarize the possible actions, gains and costs, of both the user acting as a content provider and the network, which hosts the user generated content.

1. The user faces the dilemma of either to incur the cost of requesting content identification (*enhanced* service), in addition to the payment to the network for participating in the UGC service (*basic* service), or not. The user that does not request content identification, takes the risk that such content may be identified as infringing, in which case the user may be held liable, incurring possible fines. Let the user's total cost be referred to as the compensation to the network, κ, if it encompasses the cost for content identification, or κ', if it does not, where $\kappa' < \kappa$.

2. The network will receive compensation κ or κ'. In either case, it may subsequently request the copyright broker to check the uploaded content through the broker's content identification technology framework, which is expected to detect in most cases whether a content is infringing. Thus to check content quality q, the network undertakes an additional cost. If it decides not to check quality of content (refer to unchecked quality as q'), it saves money but it risks that the content may be infringing and

---

[3]A copyright broker is an entity, which maintains a list of contents identified as *new* and uses effective content identification technology to locate possible infringing user-uploaded audio and video content.

thus may result in even greater costs in terms of fines. Thus the network's total cost for hosting this request is $c(q)$, which includes the content identification costs if it checks content quality. Otherwise, the network's total cost for unchecked content quality is $c(q')$, and $c(q') < c(q)$.

3. The user experience could be quantified in terms of satisfaction, where a content identified as non-infringing, i.e., of quality q, results in satisfaction $\pi(q)$. On the other hand, a content identified as unchecked (and possibly infringing), i.e., of quality q', results in satisfaction $\pi(q')$, where $\pi(q') < \pi(q)$, since a non-infringing content is distributed but an identified infringing content may not, and furthermore it may result in suits and fines if the user is held liable for this infringment.

We summarize the possible actions of the user and the network, for each iteration of the game whether the game consists of one single interaction or an infinite number of interactions, as well as their corresponding payoffs, in Table 2.6.

Table 2.6: User-network interaction scenario for a UGC service

	Network Checks Content	Network Does Not Check Content
User Requests Content Identification	$\pi(q) - \kappa,\ \kappa - c(q)$	$\pi(q') - \kappa,\ \kappa - c(q')$
User Does Not Request Content Identification	$\pi(q) - \kappa',\ \kappa' - c(q)$	$\pi(q') - \kappa',\ \kappa' - c(q')$

2.4.2   Incentives, assumptions and requirements

Investigating the interaction between the user and the network, we seek to identify the incentives for each entity to make a particular decision on what action or actions to take. We will investigate decisions that result in a cooperative behavior and are such that both entities are satisfied. The incentive to each participant is given by a function usually representing the participant's payoff resulting from a paticular decision.

Considering the above-mentioned scenario of user-network interaction, the payoffs of the user and the network are the following:

- The payoff of the user is the difference between the perceived *satisfaction*, which is considered to be a function of content *quality*, i.e., unchecked (possibly infringing) or checked and non-infringing, and the *compensation* offered by the user to the network, which includes the payment to participate in either *basic* UGC service or *enhanced*.

- The payoff of the network is basically the profit of the network, represented as the difference between the compensation received from the user for the *basic* or *enhanced* service, and the total cost for hosting this request, which may or may not include the cost to check whether content is infringing.

Both payoffs are based on the fact that the compensation offered by the user to the network and the satisfaction in the user payoff, as well as the total cost for hosting the request by the network are comparable. In the user payoff, the compensation and satisfaction may be measured by similar *units of satisfaction*, i.e., the more compensation given the less the *units of satisfaction* for the user, and the more quality received the more the *units of satisfaction* for the user. In the network payoff, both the cost for identifying and distributing content and the compensation received may be expressed in terms of monetary amounts. Again we may view the comparison in terms of *units of satisfaction*, i.e., the more compensation received the more the *units of satisfaction* for the network, whereas the more the cost, the less the *units of satisfaction* for the network.

In order to explore the realization of the above incentives, we proceed to model the user-network interaction as a game, based on the following *assumptions*:

Assumption 1. The two players, the user and the network, are heterogeneous, aiming at different payoffs. This assumption is realistic given the diverse nature of a network and a user.

Assumption 2. The modelling of the game assumes a complete game, i.e., one in which players are aware of the available actions and corresponding payoffs of their opponents, but of imperfect information, since a player makes decisions without having knowledge of his/her opponent's decision in the same round of the game.

Assumption 3. At any time the sum of the payoffs of the two players is a constant value, not equal to zero, i.e., a general sum game model, as opposed to no player winning as in a zero-sum game. This is a reasonable assumption since we aim to motivate a cooperative interaction.

Assumption 4. Both the user and the network have non-negative payoff functions (same reasoning as Assumption 3).

Assumption 4 results in the following requirements:

Req. 1 For the user to participate in the interaction, the satisfaction $\pi(q)$ expected to be received should be greater than the compensation κ offered to the network. This holds also for minimum satisfaction $\pi(q')$. The difference, however, between $\pi(q')$ and compensation for *basic* service κ', is minimum, whereas the difference between $\pi(q)$ and κ can be much higher, since the user wants to pay less and receive more. Given that perceived satisfaction and compensation are measurable in

comparable units, the user *plays* the game without risking to have a
negative payoff, satisfying Assumption 4.

Req. 2 For the network to participate in the interaction, we require that the
network's cost $c(q)$ from hosting the UGC service is less than the
compensation κ offered by the user by an amount $\epsilon > 0$, such that
the network may gain at least marginal profit from this interaction.
This holds for both kinds of services, i.e., when the user chooses to
give a compensation κ' for *basic service* and when the user chooses
to give a compensation κ for *enhanced* service. The network accepts
to participate in the interaction in a riskless manner, i.e., only if the
range of possible compensations exceeds the range of possible costs,
i.e., $\kappa - \kappa' > c(q) - c(q')$. Given that the compensation and cost of
supporting a requested quality are measurable in comparable units,
the network *plays* the game without risking to have a negative payoff,
satisfying Assumption 4.

Req. 3 Given marginal profit $\epsilon > 0$, it is better for a network not to reject
the user's content, but accept it and decide whether to cooperate, i.e.,
check the content, or defect, i.e., not check the content. On the other
hand, we also allow the user to be able to defect from cooperation, by
not requesting to identify the content uploaded, thus ending up giving
a lower amount of compensation to the network.

2.4.3 No past or future consideration: One-shot user-network interaction

In this section, we examine the interaction as a one-shot game, i.e., such that
it is not affected by outcomes of previous interactions and that it does not
affect any future interactions between the two entities. Given this setup we
will analyze whether the players are motivated to cooperate.

Consider a game between 2 heterogeneous players, the user and the network, interacting as follows.

Definition 2.4.1 (User-Network Interaction game). *Let the user choose between compensations κ and κ' to offer to the network for participating in a
UGC service, where κ' is the cost to participate in the* basic *service, and κ is
the cost for the* enhanced *service. The network receives the compensation κ or
κ', and decides whether to check for infringment, with total cost for hosting
the request equal to $c(q)$, or not to check for infringment with total cost equal
to $c(q')$, where $c(q') < c(q)$; q represents the content quality, i.e., quality is
represented by q if checked or by q' if not checked. The user perceives a certain satisfaction from the network's action, specifically $\pi(q)$ for distribution
of content identified as non-infringing, or $\pi(q')$ for content identified as infringing and not distributed, where $\pi(q') < \pi(q)$; in this case content quality
is represented by q if content is treated as non-infringing, or by q' if content*

is treated as infringing. Given the user and network choice of actions, both players aim to maximize their payoffs.

Table 2.6, illustrates the payoffs corresponding to each set of actions for the user and the network in the user-network interaction game.

Equivalence to Prisoner's Dilemma

Considering Definition 2.4.1 for the User-Network Interaction game, we prove next that this case is equivalent to a Prisoner's Dilemma type of game (Definition 2.3.1) by making use of Table 2.7, which illustrates the mapping between the user and network payoffs in a User-Network Interaction game and the payoffs in a Prisoner's Dilemma type of game.

Table 2.7: Mapping between user and network payoffs in Prisoner's Dilemma game

	User Payoffs (j=1)	Network Payoffs (j=2)
A_j	$\pi(q) - \kappa'$	$\kappa - c(q')$
B_j	$\pi(q) - \kappa$	$\kappa - c(q)$
C_j	$\pi(q') - \kappa'$	$\kappa' - c(q')$
D_j	$\pi(q') - \kappa$	$\kappa' - c(q)$

Proposition 2.4.1. *The User-Network Interaction game (Definition 2.4.1) is equivalent to a Prisoner's Dilemma game.*

Proof. By Definition 2.4.1 we immediately conclude that:

Observation 1. *There are two possible actions for the user: (i) to cooperate, i.e., offer compensation for enhanced service, and (ii) to defect from cooperation, i.e., offer compensation that includes payment for participation in basic service. Similarly, there are two possible actions for the network: (i) to cooperate, i.e., check uploaded content for infringment, and (ii) to defect from cooperation, i.e., not check uploaded content for infringment.*

Observation 1 combined with Definition 2.3.1 implies that the actions of the players in the user-network interaction game match the actions of the players of a Prisoner's Dilemma type of game. In particular, Table 2.7 maps each player's payoffs, of Table 2.6 to actions A_j, B_j, C_j, D_j, where $j \in \{1,2\}$, as defined in Definition 2.3.1. We proceed to prove:

Lemma 2.4.1. *Set A_j, B_j, C_j, D_j according to Table 2.7. Then it holds that: $A_j > B_j > C_j > D_j$ for each $j \in \{1,2\}$.*

Proof. The User-Network Interaction game satisfies condition 1 of Definition 2.3.1.

Examining the user, we verify straightforward that $\pi(q) - \kappa' > \pi(q) - \kappa$, thus $A_1 > B_1$, and that $\pi(q') - \kappa' > \pi(q') - \kappa$, thus $C_1 > D_1$, since $\kappa > \kappa'$. Given Assumption 4 and Req. 1, then $\pi(q) - \pi(q') > \kappa - \kappa'$, and it holds that $B_1 > C_1$.

Examining the network, we verify straightforward that $\kappa - c(q') > \kappa - c(q)$, thus $A_2 > B_2$, and that $\kappa' - c(q') > \kappa' - c(q)$, since $c(q) > c(q')$, thus $C_2 > D_2$. Assuming that the network accepts to participate in the interaction in a riskless manner, i.e., only if the range of possible compensations exceeds the range of possible costs (Assumption 4, Req. 2), then $\kappa - \kappa' > c(q) - c(q')$, and $B_2 > C_2$. $\qquad\square$

We now proceed to prove that:

Lemma 2.4.2. *The User-Network Interaction game, satisfies condition 2 of Definition 2.3.1.*

Proof. To prove the claim we must prove that the reward for cooperation is greater than the payoff for the described situation, i.e., for each player it must hold that $B_j > \frac{A_j + D_j}{2}$.

Considering Assumption 4, i.e., the payoffs of both the user and the network are non-negative, then $(\pi(q) - \kappa) > 0$ and $(\kappa - c(q)) > 0$. Combining, we have $\pi(q) > \kappa > c(q)$ and we consider this relation in the following:

For the user,

$$\pi(q) - \kappa > \frac{\pi(q) - \kappa' + \pi(q') - \kappa}{2} = \frac{\pi(q) - \kappa}{2} + \frac{\pi(q') - \kappa'}{2}, \qquad (2.1)$$

since $\pi(q) > \pi(q')$ and $\kappa > \kappa'$.

For the network,

$$\kappa - c(q) > \frac{\kappa - c(q') + \kappa' - c(q)}{2} = \frac{\kappa - c(q)}{2} + \frac{\kappa' - c(q')}{2}. \qquad (2.2)$$

$\qquad\square$

Observation 1, Lemma 2.4.1 and Lemma 2.4.2 together complete the proof of Proposition 2.4.1. $\qquad\square$

The decision of each player in the User-Network Interaction game is based on the following reasoning [5]: If the opponent cooperates, defect to maximize payoff (A_j *in Table 2.7*); if the opponent defects, defect (*payoff C_j instead of payoff D_j*). This reasoning immediately implies:

Corollary 2.4.1. *[5] In a one-shot Prisoner's Dilemma game, a best response strategy of both players is to defect from cooperation.*

Proposition 2.4.1 combined together with Corollary 2.4.1 immediately implies:

Corollary 2.4.2. *In the one-shot User-Network Interaction game, a best response strategy of both players is to cheat.*

The above discussion leads to the conclusion that cooperation between the user and the network is not supported. Hence, we next consider how past and future interactions affect and are affected by the present interaction, and whether cooperation can be better motivated. Note that in realistic scenarios it is common that previous outcomes of the interaction affect the behavior of the two entities in a present interaction. Therefore, we introduce in Section 2.4.4, an improved model of the interaction, which takes into account past interactions and how a decision could result in the better future outcome. In particular, we present a repeated game model that captures the case where players have a one-unit window of the previous history of the game, i.e., they know their opponent's last action. Thus, we investigate how the behavior of both the user and the network changes if we consider a repeated interaction between the two entities. The User-Network Interaction is modelled as a repeated game with infinite horizon [8], with the payoffs for each period as indicated in Table 2.6.

2.4.4 Considering past and future: Repeated user-network interaction

The interactions between networks and users are commonly not one-shot but re-occuring. In such relationships, the players do not only seek the immediate maximization of payoffs but instead the long-run optimal solution. Such situations are modelled in game theory by repeated game models. There are two kinds of repeated game models: the finite horizon repeated games and the infinite horizon repeated games, which are actually models of games of unknown length [8, Chapter 11]. In the given illustrative scenario we consider the user-network interaction model as an infinite horizon repeated game, since the users may keep requesting new UGC sessions from the networks for the particular service, but the number of such requests is not known. In such a scenario users and networks may choose between a number of strategies. Once chosen, we assume that the players stick to that strategy until the end of the game.

The Grim Trigger strategy, previously defined, may be used by the user in the user-network interaction game, such that if the user is not satisfied in one stage, i.e., the network does not provide the service promised, in the next stage the user may *punish* the network by leaving the relationship forever (e.g., stop interacting with the specific network for subsequent requests of the particular service). Therefore, the user employs a Grim Trigger strategy to elicit cooperative performance from the network and the loss of the relationship is costly to the network because it has a negative impact on the user-network relationship, since according to the user's strategy, if the user is not satisfied with the provided service, the user leaves the relationship, and the network loses its customer. We use this strategy for the user initially, and subsequently we will also discuss and demonstrate variations of this strategy, where the punishment part of the Grim Trigger strategy is modified. On the other hand, for the network, a sensible strategy that can be used to elicit cooperative behavior from the user is the Tit-for-Tat, since the network cannot leave a user-network relationship, it may punish a defect action by the user with a corresponding defect action, encouraging the user to keep to cooperation.

To illustrate how different strategies influence the game, in addition to the grim strategy for the user and the Tit-for-Tat strategy for the network, which are both cooperative in nature, we define two more possible defecting strategies, one for the user, the *Cheat-and-Leave* strategy, and one for the network, the *Cheat-and-Return* strategy. In the *Cheat-and-Leave* strategy, the user leaves after defecting from cooperation, i.e., does not continue interaction with the particular network in order to avoid any punishment for defecting. In the *Cheat-and-Return* strategy, the opportunity is given to the network to defect from cooperation and not check the content, and since in reality it cannot leave the user-network relationship (if the user selects to interact with the particular network), the network returns to the interaction and accepts the user's punishment, if any. Thus, we study game profiles where the strategy each player employs can purposely cooperate or purposely defect. Once we establish that the most profitable profile for both players is one where both players employ cooperative strategies, we discuss the idea of punishment and we modify the user's cooperative strategy. In particular, for the Grim Trigger strategy, in terms of the punishment part of the strategy, we consider two variations, one where the punishment is minimum, i.e., if the network defects from cooperation, it does not involve the user leaving the relationship forever, but leaving for only one period and returning in the subsequent interaction period (the *Leave-and-Return* strategy), and one where the punishment can vary according to the network's behavior (the *Adaptive-Return* strategy), which considers an enhanced state for the user device, and we provide a comparative analysis. Next, we offer definitions for the user and network defecting strategies.

Definition 2.4.2 (Cheat-and-Leave Strategy). *When the user cooperates and then defects from cooperation in a random period, immediately leaving in the next period to avoid punishment, the strategy is referred to as the* cheat-and-leave strategy.

Definition 2.4.3 (Cheat-and-Return Strategy). *When the network cooperates and then defects from cooperation in a random period, immediately returning to cooperation in the next period, the strategy is referred to as the* cheat-and-return strategy.

We are now ready to introduce a repeated game model of the user-network interaction when the history is taken into account in the decisions of the entities:

Definition 2.4.4 (Repeated User-Network Interaction Game). *Consider a game with infinite repetitions of the one shot User-Network Interaction game with one additional action available to the user: leaving the interaction[4]. Let the payoffs from each iteration be equal to the payoffs from the one-shot User-Network Interaction game, and in the case the user leaves, let the payoff in that iteration to both players be equal to zero. Then, the game is referred to as* Repeated User-Network Interaction Game.

Because of the unknown number of iterations, the PV of each player is calculated after a history, i.e., record of all past actions that the players made [8, Chapter 11], to be able to evaluate each available action in the remaining game. The PV makes use of the idea of a discount factor $\delta = \frac{1-p}{1+r}$, where $p \in [0, 1]$ is the probability of termination of the interaction, and r is the rate of satisfaction gain of continuing cooperation in the next period. Thus we may consider the cumulative payoffs for each player from the repeated interaction. Next we will show how an equilibrium exists when the user and the network employ strategies that are cooperative in nature.

User and Network Strategies Let the user have a choice between a cooperate and a defect strategy: (i) the *grim* strategy, i.e., offer a compensation to him if degradation is perceived, then leave the relationship forever, and (ii) the *cheat-and-leave* strategy. Let the network have the same kind of choice, where the cooperate and defect strategies are: (a) the *Tit-for-Tat* strategy, i.e., mimic the actions of its opponent, and (b) the *cheat-and-return* strategy.

When both players select their cooperate strategy and neither of the two players defects from cooperation, the sequence of game profiles (we will refer to this simply as profile from now on) is one of cooperation defined more formally next:

Definition 2.4.5 (Conditional-Cooperation Profile). *When the user employs the grim strategy ((i) from paragraph* User and Network Strategies*) and the network employs the Tit-for-Tat strategy ((a) from from paragraph* User and Network Strategies*), the profile of the repeated game is referred to as* conditional-cooperation profile of the game.

The following theorem states that the *conditional-cooperation* profile strategies provide a better response to the alternative strategies: *cheat-and-leave* ((ii) from from paragraph *User and Network Strategies*) for the user and *cheat-and-return* ((b) from from paragraph *User and Network Strategies*) for the network.

Theorem 2.4.1. *In the repeated user-network interaction game, assume $\delta > \frac{c(q)-c(q')}{\kappa-c(q')}$ and $\delta > \frac{\kappa-\kappa'}{\pi(q)-\kappa'}$. Then, the conditional-cooperation profile strategies result in higher payoffs than the* cheat-and-leave *and* cheat-and-return *strategies.*

---

[4]It is logical to assume that the user can switch to a different network if dissatisfied, whereas a network cannot leave the interaction once it decides to participate in it.

Proof. We assume a history of cooperative moves in the past. We compute the PVs of both the user and the network and after comparing them we conclude that the conditional-cooperation profile strategies are more motivated than the alternative *cheat-and-leave* and *cheat-and-return* strategies, under certain derived conditions.

1. Assume first that the user plays the grim strategy. Considering the network's strategies it could either play the Tit-for-Tat strategy, i.e., cooperate in the current period, or play the cheat-and-return strategy, where it may defect from cooperation.

If the network cooperates, then:

$$PV_{coop}^{net} = \frac{\kappa - c(q)}{1 - \delta}.$$

If the network defects, then:

$$PV_{def}^{net} = \kappa - c(q') + \frac{\delta \cdot 0}{1 - \delta}.$$

For the network to be motivated to cooperate, its PV in case of cooperation must be preferrable than its PV in case of defecting. Thus:

$$PV_{coop}^{net} > PV_{def}^{net} = \frac{\kappa - c(q)}{1 - \delta} > \kappa - c(q') + \frac{\delta \cdot 0}{1 - \delta}.$$

If the user plays the grim strategy, the network is motivated to cooperate when $\delta > \frac{c(q) - c(q')}{\kappa - c(q')}$.

2. Assume now that the user plays the cheat-and-leave strategy. Considering the network's possible strategies, it could either cooperate or defect from cooperation in the current period.

If the network cooperates, then:

$$PV_{coop}^{net} = \kappa' - c(q) + \frac{\delta \cdot 0}{1 - \delta}.$$

If the network defects, then:

$$PV_{def}^{net} = \kappa' - c(q') + \frac{\delta \cdot 0}{1 - \delta}.$$

If the user plays the cheat-and-leave strategy, the network is not motivated to cooperate since $PV_{def}^{net} > PV_{coop}^{net}$.

3. Assume now that the network plays the Tit-for-Tat strategy. If the user plays the grim strategy, the network will cooperate in the current period whereas if the user plays the cheat-and-leave strategy, the network may defect from cooperation.

If the user cooperates, then:

$$PV_{coop}^{user} = \frac{\pi(q) - \kappa}{1 - \delta}.$$

If the user defects, then:

$$PV_{def}^{user} = \pi(q) - \kappa' + \frac{\delta \cdot 0}{1 - \delta}.$$

For the user to be motivated to cooperate, its PV in case of cooperation must be preferrable than its PV in case of defecting. Thus:

$$PV_{coop}^{user} > PV_{def}^{user} = \frac{\pi(q) - \kappa}{1 - \delta} > \pi(q) - \kappa' + \frac{\delta \cdot 0}{1 - \delta}.$$

If the network cooperates, the user is motivated to cooperate when $\delta > \frac{\kappa - \kappa'}{\pi(q) - \kappa'}$.

4. Assume finally that the network plays the cheat-and-return strategy. Considering the user's possible strategies, it could either cooperate or defect in the current period.

If the user cooperates, then:

$$PV_{coop}^{user} = \pi(q') - \kappa + \frac{\delta \cdot 0}{1 - \delta}.$$

If the user defects, then:

$$PV_{def}^{user} = \pi(q') - \kappa' + \frac{\delta \cdot 0}{1 - \delta}.$$

If the network plays the cheat-and-return strategy, the user is not motivated to cooperate since $PV_{def}^{user} > PV_{coop}^{user}$.

It follows that the conditional cooperation profile is motivated when $\delta > \frac{c(q) - c(q')}{\kappa - c(q')}$ and $\delta > \frac{\kappa - \kappa'}{\pi(q) - \kappa'}$. ☐

Equilibria

Since in such an infinitely repeated game we have an infinite number of decision nodes, we describe decision nodes in terms of *histories*, i.e., records of all past actions that the players took [6], thus a history corresponds to a path to a particular decision node in the infinitely repeated game tree. When a strategy instructs a player to play the best response to the opponent's strategy after every history, i.e., giving the player a higher payoff than any other action available after each particular history, it is called a *subgame perfect strategy* [8]. To have a subgame perfect strategy, then we must show that for every possible iteration of the game, the current action results in the highest payoff, against all possible actions of the opponent player. In the repeated User-Network Interaction game we consider a history of cooperation from both players, and hence the best response strategy of each player against all possible actions of the opponent in the current period, is given in terms of PV, considering such a cooperative history. When all players play their subgame perfect strategies, then we have an equilibrium in the repeated game, known

as a *subgame perfect equilibrium* [8]. To prove a subgame perfect equilibrium
for the *conditional-cooperation* profile, we must show that the payoff gain from
defecting from cooperation now, i.e., the temptation in the current period, is
less than the difference between the cooperative reward and the punishment
for defecting, starting from the next period of the game and lasting forever.
This technique is based on the idea of the *single-deviation principle*, i.e., at
any stage allow one of the players to change his action. This technique may
be applied to games where previous histories of all players are known, players
move simultaneously and strategies prescribe the same behavior and payoffs
in all stages. Since our game satisfies these conditions, consider the following
theorem:

Theorem 2.4.2. *In the repeated user-network interaction game, assume*
$\delta > \frac{\kappa - \kappa'}{\pi(q) - \kappa'}$ *and* $\delta > \frac{c(q) - c(q')}{\kappa - c(q')}$. *Then, the conditional-cooperation profile is*
a subgame perfect equilibrium.

Proof. For the user:

$$Temptation_{now} = (\pi(q) - \kappa') - (\pi(q) - \kappa) = \kappa - \kappa'$$

$$Reward_{forever} = \frac{\pi(q) - \kappa}{1 - \delta}$$

$$Punishment_{forever} = \frac{0}{1 - \delta} = 0.$$

The reward and punishment are considered from the next period, therefore
we discount by δ, and we have:

$$Temptation_{now} < \delta(Reward_{forever} - Punishment_{forever})$$

$$\kappa - \kappa' < \delta(\frac{\pi(q) - \kappa}{1 - \delta} - 0)$$

$$\delta > \frac{\kappa - \kappa'}{\pi(q) - \kappa'}.$$

For the network:

$$Temptation_{now} = (\kappa - c(q')) - (\kappa - c(q)) = c(q) - c(q')$$

$$Reward_{forever} = \frac{\kappa - c(q)}{1 - \delta}$$

$$Punishment_{forever} = \frac{0}{1 - \delta} = 0.$$

Similarly to the user calculations, the reward and punishment are considered from the next period, therefore we discount by δ, and we have:

$$Temptation_{now} < \delta(Reward_{forever} - Punishment_{forever})$$

$$c(q) - c(q') < \delta(\frac{\kappa - c(q)}{1 - \delta} - 0)$$

$$\delta > \frac{c(q) - c(q')}{\kappa - c(q')}.$$

Thus, the conditional-cooperation profile is a subgame perfect equilibrium.

□

A more practical user strategy

Thus, we have shown that the most profitable choice of strategies for both the players, are strategies that are cooperative in nature, and in particular the *conditional-cooperation* profile, which is the game profile where the user employs the Grim Trigger strategy and the network employs the Tit-for-Tat strategy. Next, we try to improve upon this result by modifying the user's strategy and in particular, the punishment part of the Grim Trigger strategy. The Grim Trigger strategy is a cooperative strategy in nature but it employs the threat that if the opponent defects then the player employing the Grim Trigger strategy will leave the interaction forever. We consider that in the long-term, it is not efficient for a user to quit interacting with a particular network forever. Therefore, we revisit the user-network interaction but in this case, we replace the Grim Trigger strategy with a variation, which punishes the opponent with only one period instead of forever, i.e., if the network defects the user will leave the relationship for one period and once the period is over, return to the relatiosnhip and assume cooperative behavior. Once we study this modification, we will discuss a more efficient modification in terms of eliciting cooperative behavior, where the punishment is modified according to the network's past behavior in an adaptive manner. We begin by analyzing the *one-period punishment* modification first and then analyze the adaptive case.

Let the repeated User-Network Interaction game be modified as follows: the user may employ a strategy such that the punishment imposed to the network for defecting lasts only for one period; namely, let the user be allowed to employ the *leave-and-return* strategy as defined next:

Definition 2.4.6 (Leave-and-Return Strategy). *When the user cooperates as long as the network cooperates, and leaves for one period in case the network defects, returning in the subsequent period to cooperate again, the strategy employed is referred to as the* leave-and-return *strategy.*

Based on the newly defined strategy, the *conditional-cooperation* profile is set into a new game profile, the *one-period punishment* profile, where the Grim Trigger strategy is replaced with its modified version, the leave-and-return strategy.

Definition 2.4.7 (One-Period-Punishment Profile). *When the user employs the leave-and-return strategy and the network employs the Tit-for-Tat strategy,* *the profile of the repeated game is referred to as* one-period-punishment *profile of the game.*

It has been proven in [9], that the conditions to sustain cooperation with Grim Trigger strategies, which are the stricter strategies that may be employed in a repeated Prisoner's Dilemma, are necessary conditions for the possibility of any form of conditional cooperation. That is, a Grim Trigger strategy can sustain cooperation in the iterated Prisoner's Dilemma under the least favorable circumstances of any strategy that can sustain cooperation.

Motivated by the result in [9], we show that it is easier to impose cooperation in our game under the *conditional-cooperation* profile, than under the *one-period-punishment* profile.

Theorem 2.4.3. *Assume that $\delta > \frac{c(q)-c(q')}{\kappa-c(q')}$ and $\delta > \frac{\kappa-\kappa'}{\pi(q)-\kappa'}$ in the repeated User-Network Interaction game. Then, the conditional-cooperation profile motivates cooperation of the players. These conditions on δ are also necessary to motivate cooperation in the one-period-punishment profile.*

Proof. Given a history of cooperation, we seek the values of δ that can motivate cooperation under the *one-period-punishment* profile.

Assume first that the user cooperates in the current period. Then, the network has two options: to cooperate or to defect from cooperation.

If the network cooperates, then:

$$PV_{coop}^{net} = \kappa - c(q) + \delta \cdot (\kappa - c(q)) + \frac{\delta^2 \cdot (\kappa - c(q))}{1 - \delta}.$$

If the network defects, then:

$$PV_{def}^{net} = \kappa - c(q') + \delta \cdot 0 + \frac{\delta^2 \cdot (\kappa - c(q))}{1 - \delta}.$$

In order for cooperation to be motivated, it must be that:

$$PV_{coop}^{net} > PV_{def}^{net} =$$

$$\kappa - c(q) + \delta \cdot (\kappa - c(q)) + \frac{\delta^2 \cdot (\kappa - c(q))}{1 - \delta} >$$

$$\kappa - c(q') + \delta \cdot 0 + \frac{\delta^2 \cdot (\kappa - c(q))}{1 - \delta}.$$

Simplifying, we get $\delta > \frac{c(q)-c(q')}{\kappa-c(q)}$.

Now, assume the network cooperates in the current period. Then, the user has two options: to cooperate or to defect from cooperation.

If the user cooperates, then:

$$PV_{coop}^{user} = \pi(q) - \kappa + \delta \cdot (\pi(q) - \kappa) + \frac{\delta^2 \cdot (\pi(q) - \kappa)}{1 - \delta}.$$

If the user defects, then:

$$PV_{def}^{user} = \pi(q) - \kappa' + \delta \cdot (\pi(q') - \kappa) + \frac{\delta^2 \cdot (\pi(q) - \kappa)}{1 - \delta}.$$

For cooperation to be motivated, it must be that:

$$PV_{coop}^{user} > PV_{def}^{user} =$$

$$\pi(q) - \kappa + \delta \cdot (\pi(q) - \kappa) + \frac{\delta^2 \cdot (\pi(q) - \kappa)}{1 - \delta} >$$

$$\pi(q) - \kappa' + \delta \cdot (\pi(q') - \kappa) + \frac{\delta^2 \cdot (\pi(q) - \kappa)}{1 - \delta}.$$

Simplifying, we get $\delta > \frac{\kappa - \kappa'}{\pi(q) - \pi(q')}$. □

The cooperation thresholds for both players are summarrized in Table 2.8.

Table 2.8: Cooperation thresholds

	conditional cooperation	one-period punishment
Network Cooperates if:	$\delta_{cc}^{net} > \frac{c(q) - c(q')}{\kappa - c(q')}$	$\delta_{pun}^{net} > \frac{c(q) - c(q')}{\kappa - c(q)}$
User Cooperates if:	$\delta_{cc}^{user} > \frac{\kappa - \kappa'}{\pi(q) - \kappa'}$	$\delta_{pun}^{user} > \frac{\kappa - \kappa'}{\pi(q) - \pi(q')}$

Remark 2. *It holds that $\delta_{cc}^{net} < \delta_{pun}^{net}$ since $c(q') < c(q)$, and also that $\delta_{cc}^{user} < \delta_{pun}^{user}$ since $\pi(q') - \kappa' > 0$, hence $\pi(q') > \kappa'$. Thus, both players are more motivated to cooperate under the conditional-cooperation profile.*

It appears to be easier to sustain cooperation from the user's perspective, when strategies that involve harsher punishments are used, e.g., the grim strategy. When the user employs the grim strategy, the network is motivated to cooperate, if $\delta_{cc}^{net} > \frac{c(q) - c(q')}{\kappa - c(q')}$, whereas, if the user employs the one-period punishment strategy, the network is motivated into cooperation if $\delta_{pun}^{net} > \frac{c(q) - c(q')}{\kappa - c(q)}$. The two values for δ are similar with the only difference between the two results to be the second term in the denominator, $c(q)$ instead of $c(q')$. Therefore, the second result is always greater than the first since $c(q) > c(q')$, and hence the first result (grim strategy) is easier to motivate.

For the user, cooperation is also motivated more easily with the conditional cooperation profile since the two values of δ, i.e., $\delta_{cc}^{user} > \frac{\kappa - \kappa'}{\pi(q) - \kappa'}$ and $\delta_{pun}^{user} > \frac{\kappa - \kappa'}{\pi(q) - \pi(q')}$, differ only in the second term of the denominator. Given $\kappa' < \pi(q')$ according to Assumption 4 and Req. 1 of the user-network interaction, $\pi(q) - \kappa$, is greater than $\pi(q') - \kappa'$[5], then it holds that $\delta_{cc}^{user} < \delta_{pun}^{user}$, and hence the first result (grim strategy) is easier to motivate.

---

[5]As mentioned in the assumption presented for the one-shot user-network interaction, a user always prefers for the compensation to be kept as low as possible, whereas, on the contrary, satisfaction should be as high as possible.

2.4.5   The user as an adaptive entity

The repeated user-network interaction may be improved to reflect a more adaptive user strategy toward cooperation, depending on the overall behavior of the network and not just the previous interaction. To achieve this, we employ the idea of an adaptive player, such that the user's decision of which network to select considers a network's probability to defect from cooperation based on the acquired knowledge a user gains over the course of the repeated game [9]. This is expressed as a variable α (Definition 2.4.8), representing the estimated probability of the network cooperating. To achieve this, we assume that the user has an internal state, which modifies probability α after every interactive period with the network, and that α has a different value for each different network that interacts with the user.

Definition 2.4.8. *A user estimates a network's probability not to defect from cooperation, $\alpha \in [0,1]$ as follows: Given that the user has an expected satisfaction for a service request, e, such that $\pi(q)_{expected} \geq e$, the value of α at the end of an interactive period is modified according to (2.3).*

$$\alpha = \begin{cases} \alpha_{previous} + (\alpha_{previous} \cdot \frac{\pi(q)_{final}-e}{\pi(q)_{final}}), & \text{if } \pi(q)_{final} \geq e \text{ \& } \alpha_{now} \leq 1 \\ 1, & \text{if } \pi(q)_{final} \geq e \text{ \& } \alpha_{now} \geq 1 \\ \alpha_{previous} \cdot \frac{\pi(q)_{final}}{e}, & \text{otherwise .} \end{cases}$$

$$(2.3)$$

By introducing the variable $\alpha \in [0,1]$, the user considers the network's history, approaching the decision to cooperate in an adaptive manner. Also, let the initial value of α be eual to 1, i.e., $\alpha = 1$.

In Section 2.4.4 we have proposed two game profiles: in the *conditional-cooperation* profile, the user punishes the network forever if degradation is perceived even once, while in the *one-period-punishment* profile, the user punishes the network with only one period of absence, even if the network demonstrates degradation frequently. Considering the use of α, we propose a new strategy for the user to be employed as a means to interact with the network, which is not as harsh as the grim strategy employed in the conditional-cooperation game profile, but also not as lenient as the leave-and-return strategy employed in the one-period-punishment game profile.

Let the user's strategy be the following: cooperate as long as the network cooperates; if the network defects from cooperation, then leave for an x number of periods; after that, return and cooperate again. Let the number x be equal to 1 if $\alpha = 1$ or $\lceil \frac{1}{\alpha} \rceil$ otherwise; such that a network with a lower value of α suffers a separation of more periods with the user, whereas a network with a higher value for α is punished for fewer periods (minimum punishment is 1 period).

Definition 2.4.9. *In the repeated user-network interaction game, the adaptive-return strategy for the user dictates that if the network defects from* *cooperation, the user punishes the network by leaving for an x number of periods, before returning to cooperation. The value of x is a user-generated value and is defined next:*

$$x = \begin{cases} 1, & \text{if } \alpha = 1 \\ \lceil \frac{1}{\alpha} \rceil, & \text{otherwise}. \end{cases} \tag{2.4}$$

It is important that the network has a motivation to cooperate with the user, when the user employs the *adaptive-return* strategy. Given, that a Grim Trigger strategy that employs a punishment that lasts forever is the strategy that provides the strongest motivation to cooperate [9], we show that when the user employs the *adaptive-return* strategy, the network is always at least as motivated to cooperate with the user as when the user employs the one-period punishment strategy, and that further, in most cases the network is more encouraged than these minimum motivation bounds provided when the one-period punishment profile is used (in terms of δ values). Proposition 2.4.2 and Theorem 2.4.4 state and prove these claims.

Proposition 2.4.2. *Assume that* $\delta > \frac{c(q)-c(q')}{\kappa - c(q)}$ *in the repeated User-Network Interaction game. Then, when the user employs the leave-and-return strategy, i.e., when the profile of the game is the one period punishment profile, the network is motivated to cooperate. This condition on δ is also necessary to motivate cooperation by the network when the user employs the adaptive-return strategy.*

Proof. Given a history of the game where both players have cooperated in the past, and the user employs the adaptive-return strategy, the network has two options in the current period: cooperate or defect from cooperation. When the network cooperates, the PV is as follows:

$$PV^{net}_{coop} = \frac{(\kappa - c(q)) \cdot 1 - \delta^{x+1}}{1 - \delta} + \frac{\delta^{x+2} \cdot (\kappa - c(q))}{1 - \delta}.$$

The sum of a finite geometric progression is used to calculate the discounted value for the first $x + 1$ periods, i.e., the current period and the subsequent x periods for which the punishment would hold in case of defecting. If the network defects, its PV is:

$$PV^{net}_{def} = \kappa - c(q') + \frac{1 - \delta^x \cdot 0}{1 - \delta} + \frac{\delta^{x+2} \cdot (\kappa - c(q))}{1 - \delta}.$$

Consider the case that x equals to 1, i.e., the minimum number of periods that can be imposed as punishment. The conditions necessary to motivate cooperation in terms of δ when $x = 1$ are calculated next:

$$PV_{coop}^{net} > PV_{def}^{net} \;=\; \frac{(\kappa - c(q)) \cdot 1 - \delta^{x+1}}{1 - \delta} + \frac{\delta^{x+2} \cdot (\kappa - c(q))}{1 - \delta} \;>$$

$$\kappa - c(q') + \frac{1 - \delta^x \cdot 0}{1 - \delta} + \frac{\delta^{x+2} \cdot (\kappa - c(q))}{1 - \delta}$$

$$= \;\frac{1 - \delta^{x=1}}{1 - \delta} > \kappa - c(q').$$

For $x + 1$, the above result is equal to the following:

$$\frac{1 - \delta^2}{1 - \delta} > \frac{\kappa - c(q')}{\kappa - c(q)}.$$

Therefore,

$$\delta > \frac{c(q) - c(q')}{\kappa - c(q)}.$$

$\square$

Remark 3. *Proposition 2.4.2 proves that, in terms of δ and when the value of x is set to 1, the motivation of the network to cooperate with the user, when the user employs the adaptive-return strategy, is equal to the motivation of the network to cooperate with the user employing the leave-and-return strategy.*

Theorem 2.4.4. *When the user employs the adaptive-return strategy, the values of δ, above in which the network is motivated to cooperate, decrease as the values of x, the number of punishment periods imposed by the user, increase, i.e., as the punishment becomes harsher, the network is more motivated to cooperate.*

Proof. According to Proposition 2.4.2, when $x + 1$, the network is motivated to cooperate for $\delta > \frac{c(q) - c(q')}{\kappa - c(q)}$.

Let $x = 2$, then:

$$\frac{1 - \delta^3}{1 - \delta} > \frac{\kappa - c(q')}{\kappa - c(q)}.$$

Simplifying,

$$\delta + \delta^2 > \frac{c(q) - c(q')}{\kappa - c(q)}, \delta > \frac{c(q) - c(q')}{\kappa - c(q)} - \delta^2.$$

Since $\delta \in (0, 1)$,

$$\frac{c(q) - c(q')}{\kappa - c(q)} - \delta^2 < \frac{c(q) - c(q')}{\kappa - c(q)}.$$

Thus, the network is more motivated to cooperate when $x = 2$ than when $x = 1$.

Let $x = 3$, then:

$$\frac{1 - \delta^4}{1 - \delta} > \frac{\kappa - c(q')}{\kappa - c(q)}.$$

Simplifying,

$$\delta + \delta^2 + \delta^3 > \frac{c(q) - c(q')}{\kappa - c(q)}, \delta > \frac{c(q) - c(q')}{\kappa - c(q)} - (\delta^2 + \delta^3).$$

Since $\delta \in (0,1)$ and $(\delta^2 + \delta^3) > \delta^2$

$$\frac{c(q) - c(q')}{\kappa - c(q)} - (\delta^2 + \delta^3) < \frac{c(q) - c(q')}{\kappa - c(q)} - \delta^2.$$

Thus, the network is more motivated to cooperate when $x = 3$ than when $x = 2$.

Let $x = n$, where $n > 3$, then:

$$\frac{1 - \delta^{n+1}}{1 - \delta} > \frac{\kappa - c(q')}{\kappa - c(q)}.$$

Simplifying,

$$\delta + \delta^2 + \delta^n + \cdots + \delta^n > \frac{c(q) - c(q')}{\kappa - c(q)}, \delta > \frac{c(q) - c(q')}{\kappa - c(q)} - (\delta^2 + \delta^n + \cdots + \delta^n).$$

Since $\delta \in (0,1)$ and $(\delta^2 + \delta^3 + \cdots + \delta^n) > (\delta^2 + \delta^3)$

$$\frac{c(q) - c(q')}{\kappa - c(q)} - (\delta^2 + \delta^3 + \cdots + \delta^n) < \frac{c(q) - c(q')}{\kappa - c(q)} - (\delta^2 + \delta^3).$$

Thus, the network is more motivated to cooperate when $x = n, n > 3$, than when $x = 3$.

Let $x = n + 1$, where $n > 3$, then:

$$\frac{1 - \delta^{n+2}}{1 - \delta} > \frac{\kappa - c(q')}{\kappa - c(q)}.$$

Simplifying,

$$\delta + \delta^2 + \delta^3 + \cdots + \delta^n + \delta^{n+1} > \frac{c(q) - c(q')}{\kappa - c(q)},$$

$$\delta > \frac{c(q) - c(q')}{\kappa - c(q)} - (\delta^2 + \delta^3 + \cdots + \delta^n + \delta^{n+1}).$$

Since $\delta \in (0,1)$ and $(\delta^2 + \delta^3 + \cdots + \delta^n + \delta^{n+1}) > (\delta^2 + \delta^3 + \delta^n)$

$$\frac{c(q) - c(q')}{\kappa - c(q)} - (\delta^2 + \delta^3 + \cdots + \delta^n + \delta^{n+1}) < \frac{c(q) - c(q')}{\kappa - c(q)} - (\delta^2 + \delta^3 + \delta^n).$$

Thus, the network is more motivated to cooperate when $x = n + 1$ than when $x = n, n > 3$. $\qquad\square$

Remark 4. *Theorem 2.4.4 proves that, in terms of δ, for any other value of x generated by the user, when the adaptive-return strategy is employed, the network is more motivated to cooperate with the user, than when the user employs the leave-and-return strategy when the one-period-punishment profile of the game is played.*

Corollary 2.4.3. *In terms of δ, the network is more motivated to cooperate when the user employs the grim strategy than when the user employs the adaptive-return strategy.*

Proof. Since $\delta > \frac{c(q)-c(q')}{\kappa-c(q')}$ when the punishment lasts forever, i.e., for an infinitely large number of periods, and since a punishment of n periods is always more lenient than the punishment forever, then we may deduce that $\frac{c(q)-c(q')}{\kappa-c(q')} < \frac{c(q)-c(q')}{\kappa-c(q)} - (\delta^2 + \delta^3 + \cdots + \delta^n)$. $\qquad\square$

Remark 5. *The adaptive-return strategy generates a range of punishments, which can be at least as harsh as the leave-and-return strategy (as used in the one-period-punishment profile of the game), and not as harsh as the punishment generated by the grim strategy (as used in the conditional-cooperation profile of the game).*

2.4.6 Evaluating the game

This section examines the numerical behavior of user and network strategies defined and used for the repeated user-network interaction game in Section 2.4.4. The evaluation is based on a MATLAB implementation of an iterated user-network interaction game, where all user and network strategies are played against each other multiple times in order to evaluate the behavior of each strategy in terms of payoffs. The implementation of the user-network interaction game model was based on a publicly available MATLAB implementation of the Iterated Prisoner's Dilemma Game [10], which has been extended to include all existing and proposed strategies examined in Section 2.4.4.

The implementation makes use of the following guidelines, set to reflect the analytical model of the repeated user-network interaction game. In each simulation run, both players play their strategies and get payoffs accordingly. In the first set of simulations the payoffs are the following: when the user leaves, they both get 0 in the specific period, if one defects and one cooperates, the first gets 4 and the other gets 1, if they both defect, each gets 2, and if they both cooperate each gets 3. In the second set of simulations, we investigate the behavior of the players when the difference between defecting and cooperating increases. The payoffs for the second set of simulations are the following: when the user leaves, they both get 0 in the specific period, if one defects and one cooperates, the first gets 100 and the other gets 1, if they both defect, each gets 40, and if they both cooperate each gets 60. In the third set of simulations, we investigate the behavior of the players when there is a small difference in the payoffs between the cases of both defecting versus both cooperating. The payoffs for the third set of simulations are the following: when the user leaves, they both get 0 in the specific period, if one defects and one cooperates, the first gets 90 and the other gets 10, if they both defect, each gets 50, and if they both cooperate each gets 60. We use simple numbers as payoffs to help us get some scores for different strategy combinations but these numbers follow the relationships of the payoffs as described in their general case in the repeated game model (Table 2.7). The payoffs for the different simulation sets are summarized in Table 2.9.

Table 2.9: Payoffs for the different simulation sets

Strategies	Simulation Set 1	Simulation Set 2	Simulation Set 3
User leaves	(0,0)	(0,0)	(0,0)
One defects, one cooperates	(4,1)	(100,1)	(90,10)
Both defect	(2,2)	(40,40)	(50,50)
Both cooperate	(3,3)	(60,60)	(60,60)

Furthermore, for the adaptive strategy, the value of α is randomly generated at the beginning of each simulation run but adapted according to the network's behavior during the actual simulation (since each run simulates an iterative process). A randomly generated user satisfaction and a fixed threshold of expected satisfaction are also implemented for the strategies in all simulation runs. To demonstrate the adaptivity, Figures 2.1 and 2.2 illustrate the variation of α and of $\frac{1}{\alpha}$, which is a measure of the number of periods that a punishment lasts, during its first 1,000 measurements during the simulation.

The adaptivity is clearly indicated from the variation, as well from the tendency of α toward the value of 1; this shows that this user strategy motivates the network to cooperate more, increasing α toward 1.

The variation of $\frac{1}{\alpha}$ shows that the majority of the punishments are equal or close to the value of 1. For the rest of the punishments, we observe that they may be separated in three categories: (a) the ones above 1 but still below 25, (b) the ones between 25 and 100, and (c) the ones above 100 with occurences becoming a lot less as we move from group (a) to group (c). In fact, group (c) contains 8 out of the 1,000 measurements, i.e., 0.8%.

Figure 2.3 presents a frequency distribution of $\frac{1}{\alpha}$, in order to better visualize the overall behavior of the punishments given during these first 1,000 measurements. The frequency of occurences is measured for the following intervals of punishment periods: $1 - 5$, $6 - 25$, $26 - 50$, $51 - 100$, $101 - 150$, $151 - 200$, $201 - 250$, $251 - 300$, $301 - 351$. Clearly, the network is motivated to cooperate, a fact indicated by the high frequency of occurences in the first two intervals (especially in the first interval, i.e., < 5 punishment periods, where over 90% of the measurements lie).

A randomly generated number of iterations was run for each set of simulations to get cumulative user and network payoffs for each combination of a user strategy playing against a network strategy. The user payoffs per strategy

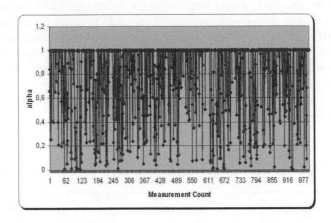

Figure 2.1: The variation of α (first 1,000 measurements).

and the network payoffs per strategy are eventually added to give the most profitable user and network strategies respectively, for the total number of iterations of a simulation run; then the average cumulative payoffs from all simulation runs are calculated. Although the number of iterations is randomly generated, we still repeat the process 100 times for each set of simulations, i.e., by randomly generating 100 different numbers of iterations, in order to include behaviors when the number of iterations is both small and large.

For the first set of simulations, i.e., with the payoffs ranging from 0 to 4, the payoffs are calculated for an average of 264.38 iterations per simulation run[6]. In both tables we see a cumulative sum of payoffs for each strategy combination. The score corresponds to either a user payoff (Table 2.10), or a network payoff (Table 2.11).

Table 2.10: User payoffs from all strategy combinations (1st simulation set)

	USER PAYOFFS	
	Network Strategies	
User Strategies	Tit-for-Tat	Cheat&Return
Grim	793.14	4.42
Cheat&Leave	6.62	4.65
Leave&Return	793.14	252.92
Adaptive-Return	793.14	355.05

The results for the first set of simulations, show that the most profitable user strategy is the *Adaptive-Return* strategy, and that the most profitable network strategy is the *Tit-for-Tat* strategy for all payoffs except for the payoff received from the combination with the user's *Cheat&Leave* strategy. This

---

[6]Minimum iterations generated: 8, maximum iterations generated: 1,259.

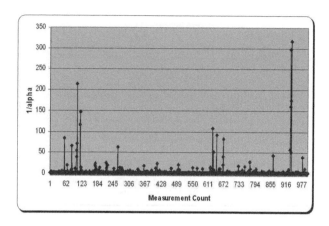

Figure 2.2: The variation of $\frac{1}{\alpha}$ (first 1,000 measurements).

Table 2.11: Network payoffs from all strategy combinations (1st simulation set)

| | NETWORK PAYOFFS | |
| | Network Strategies | |
User Strategies	Tit-for-Tat	Cheat&Return
Grim	793.14	7.42
Cheat&Leave	3.82	4.23
Leave&Return	793.14	617.21
Adaptive-Return	793.14	618.57

is seen from the corresponding cumulative payoffs for both the user and the network when playing the particular strategy, as compared to all other strategies available to the particular player. In particular, the last row in Table 2.10 shows how this strategy results in the highest payoffs for the user, no matter what the network decides to play, whereas in Table 2.11, the first column has the highest payoffs for all rows, except the second row where the results are comparable. However, the difference between the payoffs received by the network from playing either the *Tit-for-Tat* strategy or the *Cheat&Return* strategy, in combination with the user's *Cheat&Leave* strategy, is negligible. Furthermore, the combination of the two most profitable strategies in the same game profile gives the highest cumulative payoffs to both players. Based on these numerical results, we define the *Adaptive-Punishment* profile (Definition 2.4.10) for the game to consist of the *Adaptive-Return* strategy for the user and the *Tit-for-Tat* strategy for the network.

Definition 2.4.10 (Adaptive-Punishment profile). *When the user employs the* Adaptive-Return *strategy and the network employs the* Tit-for-Tat *strat-*

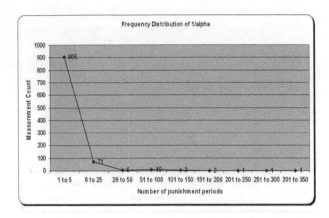

Figure 2.3: Frequency distribution of $\frac{1}{\alpha}$ (first 1,000 measurements).

egy, the profile of the repeated game is referred to as Adaptive-Punishment *profile of the game.*

For the second set of simulations, i.e., with the payoffs ranging from 0 to 100, the payoffs are calculated for an average of 240.59 iterations per simulation run[7]. As previously, we see in both tables a score for each strategy combination. The score corresponds to either a user payoff (Table 2.12), or a network payoff (Table 2.13).

Table 2.12: User payoffs from all strategy combinations (2nd simulation set)

| USER PAYOFFS | | |
| Network Strategies | | |
User Strategies	Tit-for-Tat	Cheat&Return
Grim	14435.4	47.8
Cheat&Leave	164.2	108.36
Leave&Return	14435.4	4894.76
Adaptive-Return	14435.4	4974.47

Again, we observe that for the second set of simulations, the most profitable strategy for the user is the *Adaptive-Return* strategy and the network's most profitable strategy is the *Tit-for-Tat* strategy[8]. The increase in the differences between cooperating and defecting payoffs, resulted in higher overall payoffs but has not changed the general payoff trend for the two players. In total, the highest payoffs are experienced by the players when they decide

---

[7]Minimum iterations generated: 2, maximum iterations generated: 1,124.

[8]Except for the payoff in combination with the user's *Cheat&Leave* strategy, however, the preferred profile for the game is still the *Adaptive-Punishment* profile.

Table 2.13: Network payoffs from all strategy combinations (2nd simulation set)

	NETWORK PAYOFFS	
	Network Strategies	
User Strategies	Tit-for-Tat	Cheat&Return
Grim	14435.4	146.8
Cheat&Leave	65.2	95.49
Leave&Return	14337.4	12850.4
Adaptive-Return	14435.4	12861.8

to use cooperating strategies instead of defecting; this result motivates the players to go ahead and cooperate.

For the third set of simulations, i.e., with the payoffs ranging from 0 to 90, the payoffs are calculated for an average of 268.6 iterations per simulation run[9]. As previously, we see in both tables a score for each strategy combination. The score corresponds to either a user payoff (Table 2.14), or a network payoff (Table 2.15).

Table 2.14: User payoffs from all strategy combinations (3rd simulation set)

	USER PAYOFFS	
	Network Strategies	
User Strategies	Tit-for-Tat	Cheat&Return
Grim	16416.6	62.7
Cheat&Leave	159.9	100.7
Leave&Return	16416.6	6423.6
Adaptive-Return	16416.6	6449.7

Table 2.15: Network payoffs from all strategy combinations (3rd simulation set)

	NETWORK PAYOFFS	
	Network Strategies	
User Strategies	Tit-for-Tat	Cheat&Return
Grim	16416.6	141.9
Cheat&Leave	80.7	76.7
Leave&Return	16416.6	13709.7
Adaptive-Return	16416.6	13710

Again, for the third set of simulations, the most profitable strategy for the user is the *Adaptive-Return* strategy and the network's most profitable strategy is the *Tit-for-Tat* strategy. Overall, the preferred profile for the game is still the *Adaptive-Punishment* profile. The small difference between the cases of both players defecting and of both players cooperating has not changed the

---

[9]Minimum iterations generated: 2, maximum iterations generated: 1,342.

general payoff trend for the two players. In total, the highest payoffs are experienced by the players when they decide to use cooperating strategies instead of defecting; this result continues to motivate the players toward cooperation.

Furthermore, it is worth noting that, for all simulation sets, when the user plays the *Leave&Return* strategy, the payoffs received by the user are comparable (though less) to the highest payoffs received, i.e., payoffs for employing the *Adaptive-Return* strategy. The justification for these results is the following: we have shown by Theorem 2.4.4 that the minimum conditions for the network to cooperate when the user employs the *Adaptive-Return* strategy, are also necessary for the network to cooperate when the *Leave&Return* strategy is employed. In fact, the conditions for the two strategies are at least equal, and given this, it is expected to observe payoff values that are numerically closer compared to payoff values for the other user strategies, although it appears that the *Adaptive-Return* strategy manages to achieve slightly higher payoffs than the *Leave&Return* strategy.

An additional observation is that the *Leave&Return* strategy is weaker in the case the network decides to defect, because the user's reaction is a fixed-period punishment. On the other hand, when the user employs the *Adaptive-Return* strategy, the punishment period is not fixed but adapts to the network's past behavior, appearing to consequently achieve a slight improvement in the overall user payoff.

An additional factor that we should consider when interpreting the obtained payoffs, is that each simulation investigates a single user-network interaction, thus not considering dependencies that may arise in an evaluation of multiple co-existing interactions, where a user may interact with several networks and vice versa. Such aspects are interesting and the reader is encouraged to pursue them.

Remark 6. *Regarding the payoff values of a repeated Prisoner's Dilemma type of game, it has been argued that pre-designed payoff functions that remain the same throughout the repeated game might not truly lead to autonomic players, since if the environment of the game changes during the repeated interaction, in such a way that it is not captured by the payoff functions, the players will not be able to cope [13]. Consequently, the issue of varying the payoff values has been targeted by several works; for instance, [11] investigates why the repeated Prisoner's Dilemma type of game is hardly seen in nature, arguing that the assumption of fixed payoff functions for each player is not a realistic assumption in nature, and [12] uses self adaptivity to evolve the payoff functions by evaluating feedback from the strategies and modifying the payoff functions accordingly. It is important to note that one of the factors examined in these works regarding the effectiveness of the payoffs, in continuing to motivate cooperation as the game evolves, is the spacing between the payoff values, a factor that we have adopted in the simulations presented in this book. In fact, the different simulation sets demonstrate different spacing between the players' payoffs for cooperating and defecting, showing the motivation of co-*

operation with the selected strategies throughout the evolution of the repeated game.

References

[1] ACIF Next Generation Network Project, NGN Framework Options Group, "Policy and Regulatory Considerations for New and Emerging Services," Australia, June 2006.

[2] R. M. Axelrod, "The Evolution of Cooperation," BASIC Books, New York, USA, 1984.

[3] Marty Slatnick and Geoff Parkins and Vijay Dheap, Web 2.0 meets Telecom, IBM Communications Sector, Technical Report, 2008.

[4] Bill Rosenblat, Content Identification Technologies: Business Benefits for Content Owners, GiantSteps Media Technology Strategies, Sun Microsystems, Technical Report, 2008.

[5] Graham Kendall and Xin Yao and Siang Yew Chong, World Scientific Publishing Co., The Iterated Prisoner's Dilemma: 20 Years On, Advances In Natural Computation Book Series, vol. 4, 2009.

[6] Roger B. Myerson, Game Theory: Analysis of Conflict, Harvard University Press, Cambridge, Massachusetts, 2004.

[7] Herbert Gintis, Game Theory Evolving: A Problem-Centered Introduction to Modeling Strategic Interaction, Princeton University Press, Princeton, New Jersey, 2000.

[8] Avinash Dixit and Susan Skeath, Games of Strategy, W.W. Norton & Company, New York, 1999.

[9] Rainer Hegselmann and Andreas Flache, Rational and Adaptive Playing: A Comparative Analysis for all Possible Prisoner's Dilemmas, Analyze and Kritik vol. 22, no. 1, pp. 75-97, 2000.

[10] Graeme Taylor, Iterated Prisoner's Dilemma in MATLAB, Archive for the "Game Theory" Category, http://maths.straylight.co.uk/archives/category/game-theory, March 2007.

[11] D.D.P. Johnson and P. Stopka and J. Bell, Individual variation evades the Prisoner's Dilemma, BMC Evolutionary Biology, vol. 2, no. 15, 2002.

[12] S.Y. Chong and X. Yao, Self-adapting payoff matrices in repeated interactions, IEEE Symposium on Computational Intelligence and Games, pp. 103-110, 2006.

[13] D.E. Ackley and M.L. Littman, Interactions between learning and evolution, 2nd Conference on Artificial Life, Addison-Wesley, pp. 487-509, 1991.

[14] Principles for User Generated Content Services, http://www.ugcprinciples.com, October 2007.

Chapter 3

Cooperation for two: Dealing with different types of player behavior

3.1 Introduction

In this chapter we again consider cooperation between two players. However, we deal with the case that each player may apply one of many, i.e., more than one, *facades* or types of behavior (each behavior type affects the player's preferences toward the game choices and hence the payoffs to the players). In this type of game a player does not know beforehand the type of behavior that his *multi-facade* opponent will decide to apply to a particular game instance. Although the different types of each player may be known, the fact that the choice of behavior is not known prior to the game, makes the analysis and solution of such games more challenging, since multiple information sets for each player, resulting from the different possible types of behavior, need to be taken into consideration.

We use the Bayesian type of game to model the behavior types for each player that employs more than one type of behavior, and study the resulting behaviors from the opponents. In the scenario we use to demonstrate this game, we initially use bargaining theory to *fairly* divide a certain payoff between two players. In order for this division to be fair, both players need to be truthful about the information they reveal about themselves. We consider that there are two types of players, those that are indeed truthful and those that are not. Therefore, the Bayesian game model helps us to extract conclusions on the players' types and actions. Since we make use of bargaining theory as well as Bayesian games, we next briefly introduce both of these concepts, and we move on to apply them to our illustrative networking scenario.

3.2   Cooperative behavior through bargaining

With the exception of the groundbreaking contributions of John F. Nash [1, 2], bargaining theory[1] basically evolved from the seminal paper by Ariel Rubinstein [4], which made the procedure of bargaining quite attractive, mainly due to the proposed model's simplicity and ease of understanding.

A bargaining situation is an exchange situation, in which two individuals have a common interest to *trade* but simultaneously have conflicting interests about the *price* at which to trade, because the seller would like to trade at a higher price while the buyer would like to trade at a lower price. Therefore, in a bargaining situation, the two players have a common interest to cooperate but have conflicting interests about exactly how to cooperate. On the one hand, each player would like to reach an agreement rather than disagree; on the other hand, each player wants to reach an agreement that is as favorable to that player as possible.

Therefore, a bargaining situation may be easily seen as a game situation since the outcome of bargaining depends on both players' bargaining strategies, i.e., whether or not an agreement is reached and the terms of that agreement (if one is reached) depend on both players' actions during the bargaining process. In the following paragraphs we provide brief descriptions of two well-known game models of a bargaining situation between two players, the Nash bargaining game model and the Rubinstein bargaining game model, since we will make use of these models in particular formulations in the example scenario.

The Nash bargaining game model defines a solution (known as the Nash Bargaining Solution) by a fairly simple formula, and it is applicable to a large class of bargaining situations. The Nash bargaining example is the situation where two individuals bargain over the partition of a cake of fixed size. Since the cake will be partitioned to the two players, the addition of their partitions should equal the total cake, therefore the set of non-zero partitions, which sum to the total amount is the set of possible agreements in the bargaining situation. In the case of disagreement, each player receives a penalty, which is defined according to the bargaining situation under consideration; the definition of a penalty comes from the fact that in bargaining situations the desirable outcome is agreement, thus disagreement results in a non-satisfactory payoff for the two players, which often is zero. The penalties of the two players are defined as the *disagreement point* of the game. Thus, the *useful payoff* for each player may be defined to be the player's payoff received from the received partition in case of agreement, minus the penalty that would be received in case of disagreement as defined in the *disagreement point*. The unique solution of the cake partition is therefore, the unique pair of partitions that maximizes the product of the players' *useful payoffs* and is referred to as the *Nash product* or the *Nash Bargaining Solution*.

---

[1]The text on bargaining is mainly based on [3].

The Rubinstein bargaining game is modelled as a sequential-moves game, in which the players take turns to make offers to each other until agreement is secured. This model has much intuitive appeal, because a lot of real-life negotiations are based on the idea of making offers and counter-offers. From the sequential-moves model of the bargaining process, it is easy to see that if the two negotiating players do not incur any costs or penalties for delaying the agreement decision, the solution to the bargaining game may be indeterminate, because the two players could continue to negotiate forever. Given that there is a cost to each player for delaying, then each player's bargaining power is determined by the magnitude of this cost. Consider for the Rubinstein model where, similarly to the Nash bargaining game model described above, the two players bargain over the partition of a cake of fixed size. The first player proposes a partition; if the second player accepts, agreement is reached and the game is over; otherwise, the second player proposes a different partition, and the process of alternating offers continues until an offer is accepted. However, for each additional negotiation round there is a cost to each player, i.e., the size of the cake becomes smaller. The factor by which the cake gets smaller may be different for each player, and it is referred to as the player's discount factor. The Rubinstein bargaining game model has a unique subgame-perfect equilibrium, which makes use of the fact that any offer made now by a player should be equal or greater to the discounted best value that the opponent can get in the next period.

3.3 Bayesian type of games

Another element that must be considered in bargaining situations, is truthfulness of the players. In order to motivate the two bargaining players to be truthful about their own information that may affect the bargaining process, there must exist a mechanism that can penalize a player who turns out to lie, assuming that it is detectable whether a player has lied or not. Such mechanisms exist and are called *pricing mechanisms* [5], constituting an interesting and very promising way to guarantee truthfulness of the participating entities. In the worldwide literature there is a whole research field that is focused on the development, limitations and capabilities of such pricing mechanisms; the Algorithmic Mechanism Design [5]. Successful paradigms in this context include (combinatorial) auctions [9] and task scheduling [10, 11], using techniques such as the Revelation Principle [5, 12, 13], Incentive-Compatibility [5, 12, 13], Direct-Revelation [5, 12, 13] and Vickrey-Clarke-Groves Mechanisms [14]. The algorithmic tools and theoretical knowledge that have already developed in the field of Algorithmic Mechanism Design constitute a fruitful pool for extracting algorithmic tools for enforcing players to truthfulness, through pricing mechanisms, once these tools are customized and further developed for handling the needs of any specific scenario.

In several situations where interactions occur, the interacting entities may not have complete information about each other's characteristics. The model of a Bayesian game[2] is designed to model such situations. A player's uncertainty about the opponent's characteristics, is modelled by introducing a set of possible states, i.e., probable sets of characteristics that a player may have, also known as a player's *types*. Each player assigns a probability of occurence to each of the opponent's possible *types*. Therefore, a definition of a Bayesian game is similar to the definition of a normal form game, with the additional elements of the *types* for each player and the corresponding probability of occurence, as believed by the player's opponent(s).

A Bayesian game can be modelled by introducing Nature as a player in the game[3]. Nature assigns a random variable to each player, which could take values of types for each player and associate probabilities with these types. In the course of the game, Nature randomly chooses a type for each player according to the probability distribution across each player's type space. The type of a player determines the player's payoff and the fact that a Bayesian game is one of incomplete information means that at least one player is unsure of the type and thus the payoff of another player.

In any given play of a Bayesian game, each player knows his type and does not need to plan what to do in the hypothetical event that he is of some other type. However, when determining a player's best action, he must consider what the other player(s) would do if any of the other possible types were to occur, since any player may be imperfectly informed about the current state of the game. Therefore, a Nash Equilibrium of a Bayesian game is the Nash equilibrium of the normal form game in which the set of players includes all possible types for each player, and consequently the set of actions includes all possible actions for each such state of every player considered. In brief, to reach a Nash Equilibrium in a Bayesian game, each player must choose the best action available to him, given his belief about the occurence of his opponent's types, the state of the game and the possible actions of his opponent.

3.3.1   An example of a Bayesian type of game

Let's consider an example of a two-player interaction. Suppose that the preferences of Player 1 is known to both players but the preferences of Player 2 is known only to Player 2. However, Player 1 believe that there are two possibilities regarding the preferences of Player 2, therefore, to Player 1 there are two different versions of the game that he can play with Player 2. Next we represent in normal form the two versions of the game that can result from the interaction between Player 1 and Player 2. Table 3.1 and Table 3.2 illustrate these two versions.

Given that Player 1 does not know which of the two versions of the game will be played, it becomes more difficult to decide what action to take. He needs to play according to his belief about which of the two versions of the

---

[2]This text is mainly based on [15].

[3]This approach was proposed by John C. Harsanyi in [16].

Table 3.1: Bayesian game example - version 1

	Player 2 Action 1	Player 2 Action 2
Player 1 Action 1	4, 1	3, 2
Player 1 Action 2	0, 1	1, 0

Table 3.2: Bayesian game example - version 2

	Player 2 Action 1	Player 2 Action 2
Player 1 Action 1	1, 0	0, 1
Player 1 Action 2	1, 3	1, 1

game is being played. On the other hand, Player 2, who knows his own preferences can easily make a decision since he knows the payoffs of both versions of the game, and he is the one deciding which of the two versions to play. Suppose Player 1 believes that version 1 is played. Then Player 1 should play the game accordingly. Furthermore, in analyzing such games, we should consider how important it is that Player 2 does not know which version of the game Player 1 believes is being played. Since the strategic interaction depends on which version of the game is being played, it is important to have assumptions about the preferences of the players. The more complex the game, in terms of the version of the game for each player and the number of players, the more complex the hierarchy of assumptions necessary to solve the game.

In this game there are two information sets for Player 2 and one information set for Player 1, i.e., Player 2 has 4 (2 in each version) actions to choose from whereas Player 1 has 2 actions to choose from. Thus, we can rewrite the two versions of the game as one game where Player 1 has 2 possible actions and Player 2 has 4 possible actions. Table 3.3 illustrates the new game.

Table 3.3: Bayesian game example – combined game

	Pl. 2 Action 1	Pl. 2 Action 2	Pl. 2 Action 3	Pl. 2 Action 4
Player 1 Action 1	4, 1	3, 2	1, 0	0, 1
Player 1 Action 2	0, 1	1, 0	1, 3	1, 1

Once the two versions of the game are combined into one game with more actions, it is easier to find the Nash Equilibria of the new game. First, let's consider Player 2's best responses to Player 1's actions. Player 2's best response to Player 1's Action 1 is Player 2's Action 2, and Player 2's best response to Player 1's Action 2 is Player 2's Action 3. Then, given Player 2's best responses to Player 1's actions, Player 1's best responses are the following: the best response to Player 2's Action 2 is Player 1's Action 1 and the best response to Player 2's Action 3 is again Player 1's Action 1. Hence there is one Nash Equilibrium in pure strategies, which is (Player 1's Action 1, Player

2's Action 2).

3.4   When payoffs need to be partitioned: Player truthfulness

3.4.1   Scenario overview

The scenario explores the need for synergy between two networks (in particular two Mobile Virtual Network Operators participating in the heterogeneous system) to serve a premium service user request, in order to be able to provide some additional service quality guarantees. Such cooperation between two selected networks is transparent to the user, i.e., the user offers some payment (e.g., for the premium service) without the knowledge that this payment will be partitioned between two networks. The scenario demonstrates the optimal way to partition the payment so that the two networks are motivated to cooperate. An additional issue arises in this cooperation, which is whether the cooperating networks are motivated to be truthful and how the optimal solution is affected by non-truthful behavior. We show that being truthful to support cooperation is the best mode of behavior for the participating networks in such a situation.

In a converged network, there may exist multiple Mobile Virtual Network Operators (MVNOs), each one of them interacting with the participating users through the use of a common IP-based mobile communications infrastructure, operated by a single Mobile Network Operator (MNO) who rents out resources to the MVNOs. These MVNOs use this mobile infrastructure to motivate user participation through their offered services achieving individual revenues by offering various services to these users.

In this scenario, we consider the case where enhanced quality demands by the users for a particular service may require the cooperation of two MVNOs in advance to support a particular user. The two MVNOs are selected by the use of a prioritized list based on the estimation of the satisfaction to be received by the user. Once the selection is completed the two *best* MVNOs, are encouraged to cooperate, with their revenues coming from a payment partition of the payment for the particular service, dependent on the service they provide. The payment partition is basically a configuration, pre-calculated and adopted by the two MVNOs in order to avoid either of the two MVNOs gaining bargaining advantage by handling the partition, and furthermore to ensure that the whole process is transparent to the user, obeying the user-centric paradigm characterizing converged, next generation communication networks.

Consider the case where a customer of the converged mobile communication networks makes a service request for a particular premium service with critical delay constraints for the service session. Both the preferred network and the second best network are MVNOs, both networks renting out resources from the same MNO. Further consider that it is advantageous for two MVNOs to cooperate in order to offer the user a higher guarantee of service delivery as requested. Then, to guarantee quality in terms of delay, it is beneficial for the two MVNOs to cooperate to support the particular user. The best MVNO will serve the request and simultaneously the second best MVNO will reserve resources, in order to act as a secondary network in case quality degradation is detected and/or session handoff is necessary. This cooperation enables service provision to be offered by the best MVNO and in case of need of session handoff, the session handoff is faster, enhancing service quality in terms of service continuity and handoff delay, which is a crucial aspect for real-time, critical services (e.g., medical video).

Since a network's satisfaction is represented by its revenue gain, and since two networks must cooperate for a single service, then the payment for supporting the service needs to be partitioned between them, in order for the networks to have an incentive to cooperate. Moreover, the partitioning configuration must be such that it is satisfying to both networks. Thus cooperation can be modelled as a solution to a cooperative bargaining game, which is presented next.

Next we illustrate through a representative scenario the *payment partition* as a game of bargaining between the two MVNOs, i.e., between two networks, which demonstrates the optimal way to partition a payment so that the two networks are encouraged to cooperate. Firstly, we define the payment partition as a game between the two MVNOs and we show that this is equivalent to the well-known Rubinstein bargaining game [3, Chapter 3], when the agreement is reached in the first negotiation period. Given this equivalence, an optimal solution to the Rubinstein bargaining game, would also constitute an optimal solution to the payment partition game. The resolution of the game is presented in 3.4.2.

3.4.2   Cooperative bargaining model

Let $q \in Q$ be the quality level for which the two networks negotiate. Consider the payment partition scenario, where two networks want to partition a service payment $\Pi(q)$ set by the converged platform administrator, such that the first network receives partition π_1 and the second network receives partition π_2. Let $c_i(q)$ be considering the resource reservation cost of network i, such that $c_1(q)$ is the resource reservation cost for the first network and $c_2(q)$ is the resource reservation cost for the second network.

Given the cost characteristics of network i, each network seeks a portion:

$$\pi_i(q) = c_i(q) + \phi_i(q), \tag{3.1}$$

where $\phi_i(q)$ is the actual profit of network i, such that:

$$\pi_1(q) + \pi_2(q) = \Pi(q), \tag{3.2}$$

where $\Pi(q)$ is the total payment announced by the converged platform administrator.

The networks' goal is to find the payment partition, which will maximize the value of $\phi_i(q)$, given the values of $\Pi(q)$ and $c_i(q)$. Definition 3.4.1 defines the bargaining game between the two networks:

Definition 3.4.1 (Payment-Partition game). *Fix a specific quality level $q \in Q$ such that a fixed payment Π is received. Consider a one-shot strategic game with two players corresponding to the two networks. The profiles of the game, i.e., the strategy sets of the two players, are all possible pairs (π_1, π_2), where $\pi_1, \pi_2 \in [0, \Pi]$ such that $\pi_1 + \pi_2 = \Pi$. All such pairs are called agreement profiles and define set S^a. So, $S^a = \pi_1 \times \pi_2$. In addition, there exists a so-called disagreement pair $\{s_1^d, s_2^d\}$, which corresponds to the case where the two players do not reach an agreement. So, the strategy set of the game is given by $S = S^a \bigcup \{s_1^d, s_2^d\}$. For any agreement point $s \in S^a$ the payoff $U_i(s)$, for player $i \in [2]$, is defined as follows:*

$$U_i(s) = \pi_i - c_i. \tag{3.3}$$

Otherwise,

$$U_1(s_1^d) = U_2(s_2^d) = 0. \tag{3.4}$$

This game is referred to as the payment-partition game.

The payoff $U_i(s)$ for a fixed quality level $q \in Q$ is actually, the profit $\phi_i(q)$ for player i. The aim of the game is therefore to maximize the profit by finding the optimal payment partition π_i such that when the specific resource reservation costs are subtracted, results in the highest possible profit for network i. Fact 1 states this goal more formally.

Fact 1. *Let $s^* = (\pi_1^*, \pi_2^*)$ be an optimal solution of the payment-partition game. Then $U_i(s^*) = \phi_i = \pi_i^* - c_i$, where $i \in [2]$, comprises an optimal solution of the payment partition scenario.*

Equivalence to a Rubinstein Bargaining game

Initially, we show the equivalence between the *payment-partition* game and a Rubinstein Bargaining Game, a.k.a., the basic alternating-offers game defined next according to [3]:

Definition 3.4.2 (Rubinstein Bargaining game). *Assume a game of offers and counteroffers between two players, $\pi_i^r(t)$, where $i \in \{1, 2\}$ and t indicates the time of the offer, for the partition of a cake, of initial size of Π^r. The offers continue until either agreement is reached or disagreement stops the bargaining process.*

At the end of each period without agreement, the cake is decreased by a factor of δ_i. If the bargaining procedure times out, the payoff to each player is 0. Offers can be made at time slot $t \in \mathcal{N}_0$. If the two players reach an agreement at time $t > 0$, each receives a share $\pi_i^r(t) \cdot t \cdot \delta_i$, where $\delta_i \in [0,1]$ is a player's discount factor for each negotiation period that passes without agreement being reached. The following equation gives the payment partitions of the two players:

$$\Pi^r(t) = \pi_1^r(t) \cdot t \cdot \delta_1 + \pi_2^r(t) \cdot t \cdot \delta_2. \tag{3.5}$$

$\Pi^r(t)$ decreases as time passes by and the partitions available to the two bargaining players decrease according to δ_i for each player i.

So, if agreement is reached in the first negotiation period, the payment partition is as follows:

$$\pi_1^r = \Pi^r - \pi_2^r. \tag{3.6}$$

The payment partition $\pi_i^r(t)$ of the players $1, 2 \in [2]$ if the agreement is reached in iteration t is the following:

$$\pi_1^r(t) = \Pi^r(t) - \pi_2^r(t). \tag{3.7}$$

Such a game is called a Rubinstein Bargaining game.

The cost of each player is defined by the decrease in the portion of each player, resulting in a decrease in the overall available quantity as time passes by. Thus the total quantity to be partitioned, $\Pi^r(t)$ exhibits a decrease in its value as t increases.

Proposition 3.4.1. *Fix a specific quality q. Then, the payment-partition game is equivalent to the Rubinstein Bargaining game, when the agreement is reached in the first negotiation period.*

Proof. Assuming that an agreement in the Rubinstein Bargaining game is reached in the first negotiation period $t = 1$, then the game satisfies the following:

$$\pi_1^r(1) + \pi_2^r(1) = \Pi^r(1),$$

which is a constant.

In the payment partition game, assuming an agreement profile s, we have:

$$U_1(s) + U_2(s) = \pi_1 - c_1 + \pi_2 - c_2 \quad = \Pi - c_1 - c_2,$$

since $\Pi = \pi_1 + \pi_2$ and c_1, c_2 are constants for a fixed quality level. It follows that $U_1(s) + U_2(s)$ is also constant. It follows that the *Rubinstein Bargaining* game and the *Payment Partition* game are equivalent. $\square$

We define:

Definition 3.4.3 (Optimal Payment Partition). *The optimal partition is when bargaining ends in an agreement profile that gives the highest possible payoff to each player given all possible actions taken by the opponent.*

Proposition 3.4.1 immediately implies:

Corollary 3.4.1. *Assume that agreement in a Rubinstein Bargaining game is reached in the first negotiation period, that $\Pi^r = \Pi$, and that the corresponding profile s^*, is an optimal partition for the Rubinstein game. Then, s^* is also an optimal partition for the payment-partition game.*

In the payment-partition game, unlike the Rubinstein game, we want to utilize a solution that does not have the element of time and how the bargaining is affected in sequential rounds of the game. This is because the configuration that will represent the optimal partition of the payment needs to be utilized immediately, i.e., represent the first round of negotiations. Since the Nash bargaining game [3, Chapter 2], which offers such solution, and the payment-partition game are equivalent, we utilize the solution of a Nash bargaining game in order to compute an optimal solution, i.e., a configuration, which is satisfactory for the two networks in terms of payoffs from the payment-partition game.

The solution of the Nash bargaining game, known as the *Nash Bargaining Solution*, captures such configuration, where the two bargaining game players are both satisfied. Therefore, since disagreement results in payoffs of 0, we are looking for an agreement profile $s = (\pi_1, \pi_2)$ such that the corresponding partition of the players is an optimal payment partition, i.e., the partition that best satisfies both players' objectives (Definition 3.4.3). Section 3.4.2 elaborates on resolving the payment-partition game through the use of the Nash Bargaining solution and further addresses issues of truthfulness that arise when applying such configuration.

Payment Partition based on the Nash Bargaining Solution

In Section 3.4.2, we have shown how to model cooperation between the best and second best network in order to enable service continuity during a service session. Since disagreement is not a desirable strategy for either of the two cooperating players, we may conclude that the two players will reach an agreement. Consequently, the payment must be partitioned in such a way that both the participating access networks are satisfied. To reach the optimal solution we utilize the well-known *Nash Bargaining Solution* [3, Chapter 2], which applies to Rubinstein bargaining games when the agreement is reached in the first negotiation period, and therefore to the payment-partition game, given the equivalence presented in Section 3.4.2. Since the Nash bargaining game and the payment-partition game are equivalent, we compute an optimal solution, i.e., a configuration, which is satisfactory for the two networks in terms of payoffs from the payment-partition game. The solution of the Nash bargaining game, known as the *Nash Bargaining Solution*, captures such configuration. Therefore, since disagreement results in payoffs of 0, we are looking for an agreement profile $s = (\pi_1, \pi_2)$ such that the corresponding partition of the players is an optimal payment partition, i.e., the partition that best satisfies both players' objectives.

The next Theorem proves the existence of an optimal partition of the payment between the two players, given each network's cost c_i.

Theorem 3.4.1. *There exists an optimal solution for the payment-partition game, and is given by the following:* $\pi_1 = \frac{1}{2}(\Pi + c_1 - c_2)$, $\pi_2 = \frac{1}{2}(\Pi + c_2 - c_1)$.

Proof. We consider only agreement profiles and thus refer to the partition π_i assigned to player i. In any such profile it holds that $\pi_1 + \pi_2 = \Pi$. Assuming a disagreement, this implies that cooperation fails between the two networks and the payoff gained by player i equals to $U_i(s^d) = 0$. Since in any such profile, it holds that $U_i(s^a) > 0$ it follows that the disagreement point is not an optimal solution. Since the payment-partition game is equivalent to the Nash bargaining game, a Nash Bargaining Solution (NBS) of the bargaining game is an optimal solution of the payment-partition game between two players, i.e., a partition (π_1^*, π_2^*) of an amount of goods (such as the payment). According to the NBS properties it holds that:

$$NBS = (U_1(\pi_1^*) - U_1(s^d))(U_2(\pi_2^*) - U_2(s^d))$$
$$= max(U_1(\pi_1) - U_1(s^d))(U_2(\pi_2) - U_2(s^d))$$
$$\|\leq \pi_1 \leq \|,$$
$$\pi_2 = \Pi - \pi_1.$$

Since $U_1(s^d) = 0$, $U_2(s^d) = 0$ and $\pi_2 = \Pi - \pi_1$:

$$max(\pi_1 - c_1)(\Pi - \pi_1 - c_2) =$$
$$(-2\pi_1 + \Pi - c_2 + c_1) = 0.$$

Therefore,

$$\pi_1 = \frac{1}{2}(\Pi - c_2 + c_1), \ \pi_2 = \frac{1}{2}(\Pi + c_2 - c_1). \tag{3.8}$$

□

Remark 7. *The optimal partition based on the Nash Bargaining Solution, leaves us with some important findings. Firstly, it is important to recognize that disagreement is a non-optimal solution and thus it is not preferred by the negotiators for selfish reasons, i.e., to increase their payoff. Thus, cooperation is wanted in this situation, but it is also needed in order to reach a partition agreement. There is actually an optimal way to partition the payment, however, this depends on each network's own cost of resource reservation, which is not publicly available information. This may cause each network to be tempted to lie about it, and in fact, this becomes an issue in this game, which we address in subsequent sections in this chapter.*

Additional considerations for the payment partition

The payment partition, as we have seen in the previous section, assigns a payment to each of the networks for the whole duration of the service activation. However, networks often demonstrate unplanned degradation of quality of service due to factors such as usage, or even environmental conditions that cannot be controlled but once the payment is partitioned any degradation events are not taken into account. It is possible to monitor the performance of different networks and expect such degradation events with some probability. In fact, the expectation can be different for the negotiating networks, and the question is the following: Would the optimal payment partition change if the element of degradation expectation is considered in the partitioning process? We proceed to investigate how the solution to the payment-partition behaves when we consider the existence of a constant set by the converged platform administrator representing the probability of degradation.

Theorem 3.4.2. *Assume that the converged platform administrator assigns a constant value p_i^f to network i representing the expected quality degradation, based on the particular service and current network conditions. Then, the value of the optimal solution is the same as in Theorem 3.4.1.*

Proof. Our game has the same strategy set as before. Concerning the utility functions of the players in case of disagreement, we have also $U_1(s^d) = 0$, $U_2(s^d) = 0$, $\pi_2 = \Pi - \pi_1$ as before. In case of agreement, we have in addition a constant probability p_i^f in the payoff function of each network:

$$U_i(s) = (1 - p_i^f)(\pi_i - c_i).$$

Therefore,

$$max(1 - p_1^f)(\pi_1 - c_1)(1 - p_2^f)(\Pi - \pi_1 - c_2) =$$
$$(1 - p_1^f)(1 - p_2^f)(-2\pi_1 + \Pi - c_2 + c_1) = 0.$$

The optimization shows that if constant probabilities are considered, the optimal partition is still as previously, i.e., the networks' cost is the deciding factor for the optimal solution similarly to Equation 3.8:

$$\pi_1(q) = \frac{1}{2}(\Pi - c_2 + c_1), \quad \pi_2 = \frac{1}{2}(\Pi + c_2 - c_1). \tag{3.9}$$

$\square$

Remark 8. *Note that the estimated probability that the platform administrator may consider for each participating network does not affect the optimal partition, but affects each network's payoff function, therefore a network with a high estimated probability of degradation will receive much less of a payoff than the optimal payment partition calculated according to its cost.*

3.4.3   A Bayesian form of the payment-partition game

Since the partitioning is based on each network's cost, it is required that the networks are truthful about their costs. Truthfulness is a very important consideration in cooperative situations, especially in bargaining games. The question that arises is whether it would be wise for a player to lie, considering that the player cannot be aware of who the other player is from the original set of available networks and thus cannot guess whether the other player has more or less cost, thus not being able to correctly assess the risk of such an action. Consider in our scenario that the infrastructure costs of two MVNOs renting out infrastructure from the same MNO are known, and lying over these can be easily verified, since the costs are advertised by the MNO. However, infrastructure costs are only a part of service costs, which may additionally include content costs etc., usually different for each MVNO.

A Bayesian game [7, Chapter 5] is a strategic form game with incomplete information attempting to model a player's knowledge of private information, such as privately observed costs, that the other player does not know. Therefore, in a Bayesian game, each player may have several types of behavior (with a probability of behaving according to one of these types during the game). We use the Bayesian form for the *payment-partition* game, in order to investigate the outcomes of the game, given that each network does not know whether the cost of its opponent is lower or higher than its own.

Let each network in the *payment-partition* game have two types: the *lower-cost* type (including networks of equal cost) and the *higher-cost* type. Suppose that each of the two networks has incomplete information about the other player, i.e., does not know the other player's type. Furthermore, each of the two networks assigns a probability to each of the opponent's types according to their own beliefs and evaluations. Let p_i^l be the probability according to which, network i believes that the opponent is likely to be of type *lower-cost*, and $p_i^h = (1 - p_i^l)$ be the probability according to which, network i believes that the opponent is likely to be of type *higher-cost*.

Since the two players are identical, i.e., they have the same two types and the same choice of two actions, we will only analyze network i; conclusions also hold for network j, where $i, j \in [2], i \neq j$. Therefore, network i believes that network j is of type *lower-cost* with probability p_i^l, and of type *higher-cost* with probability $1 - p_i^l$. Each network has a choice between two possible actions: to declare its own real costs (D) or to cheat (C), i.e., declare higher costs $c_i' > c_i$. The possible payoffs for network 1 are given in Table 3.4 and in Table 3.5. The general form of the payoffs for the two types appears identical here but we must consider that the possible values of the costs will be different, resulting in overall different numerical values for the payoffs of each type, even though these are calculated with the same payoff formulae (Table 3.4 and Table 3.5).

Let us consider a numerical example of the network types and their payoffs according to Table 3.4 and Table 3.5. Consider that the cost c_i equals 10 but the network is considering cheating and declaring that the cost c_i' equals 12.

Table 3.4: Network i payoffs when opponent is of type *lower-cost*

	Network j Actions	
	D	C
Network i Actions		
D	$\frac{1}{2}(\Pi + c_i - c_j)$	$\frac{1}{2}(\Pi + c_i - c_j')$
C	$\frac{1}{2}(\Pi + c_i' - c_j)$	$\frac{1}{2}(\Pi + c_i' - c_j')$

Table 3.5: Network i payoffs when opponent is of type *higher-cost*

	Network j Strategies	
	D	C
Network i Strategies		
D	$\frac{1}{2}(\Pi - c_j + c_i)$	$\frac{1}{2}(\Pi - c_j' + c_i)$
C	$\frac{1}{2}(\Pi - c_j + c_i')$	$\frac{1}{2}(\Pi - c_j' + c_i')$

Consider, that the cost c_j of the opponent is either equal to 8 or equal to 11 if the opponent is truthful, or four units higher if the opponent is cheating, i.e., c_j' is equal either to 12 or 15. Table 3.6 and Table 3.7 demonstrate these values with regards to the payoffs of network i, and we observe that cheating by either player changes the payoffs significantly, thus offering the motivation for the players to cheat.

Table 3.6: Network i payoffs when opponent is of type *lower-cost*

	Network j Actions	
	D	C
Network i Actions		
D	$\frac{1}{2}(\Pi + 10 - 8)$	$\frac{1}{2}(\Pi + 10 - 12)$
C	$\frac{1}{2}(\Pi + 12 - 8)$	$\frac{1}{2}(\Pi + 12 - 12)$

Lemma 3.4.1. *If network i believes that the probability p_i^l, i.e., that network j is of type* lower-cost, *is higher than the probability p_i^h, then it is more motivated to lie, where $i, j \in [2], i \neq j$.*

Proof. In Table 3.4, network i has higher or equal costs to network j since network j is of type *lower-cost*, thus $c_i \geq c_j$. When both players play D, i.e., they both declare their real costs, an equal or greater piece of the payment is assigned to network i, since the partition of the payment is directly proportional to the networks' costs. If network i plays C, i.e., cheats, while network j plays D, then $c_i' > c_i > c_j$, a profitable strategy for network i, since an even greater piece of the payment will be received. For the cases that network j decides to play C, then the payment partition may or may not favor network j (it depends on the actual amount of cheating, and the action of network i). If network i plays C, then it is more likely that $c_i' > c_j'$, and network i will get a greater piece, than if it plays D. $\square$

Table 3.7: Network i payoffs when opponent is of type *higher-cost*

	Network j Strategies	
	D	C
Network i Strategies		
D	$\frac{1}{2}(\Pi - 11 + 10)$	$\frac{1}{2}(\Pi - 15 + 10)$
C	$\frac{1}{2}(\Pi - 11 + 12)$	$\frac{1}{2}(\Pi - 15 + 12)$

Lemma 3.4.2. *If network i believes that the probability p_i^h, i.e., that network j is of type* higher-cost*, is higher than the probability p_i^l, then it is more motivated to lie, where $i, j \in [2], i \neq j$.*

Proof. In Table 3.5, network i has lower costs compared to network j, thus $c_i < c_j$. When both players play D, i.e., they both declare their real costs, an equal or greater piece of the payment is assigned to network j. If network i plays C, i.e., cheats then $c_i' > c_i$, so playing C will end up in a higher payoff for network i, and in case network j plays D, i may even get the bigger piece of the partition. If network j plays C, it is still better for network i to play C, since this will end up in network i receiving a greater piece than it would if it plays D when network j plays C, although, more likely, not the greater of the two pieces. □

Motivating Truthfulness

Proposition 3.4.2. *Two networks playing the Bayesian form of the payment-partition game, are not motivated to declare their real costs but instead they are motivated to cheat and declare higher costs, i.e $c_i' > c_i$, $i \in [2]$, in order to get greater payoffs.*

Proof. Straightforward by Lemma 3.4.1 and Lemma 3.4.2. □

In order to motivate the two networks to declare their real costs, there must exist a mechanism that can penalize a player who turns out to lie on its real cost, assuming that it is detectable whether a player has lied or not[4]; we refer to such mechanisms as pricing mechanisms [5].

Let the converged platform administrator be able to detect after the service session has terminated, whether either of the participating networks has lied about its costs. In order to motivate the networks to declare their real costs we introduce a *pricing mechanism*, i.e., a new variable that tunes the resulting payoffs in the payoff function of each player[5]. The pricing mechanism is a

---

[4]In fact, it is possible to estimate the costs of the two MVNOs considering that their infrastructure costs are known, content costs are usually advertised by content providers, and cost variations based on clientele and coverage can be estimated since both MVNOs are part of the same converged communication platform.

[5]A side effect for a network that decides to cheat, is that it risks not to be selected for supporting the service in the first place, since by declaring higher costs, the compensation received from the user might be affected, and subsequently, the user might not select the particular network.

post-game punishment, i.e., cheating in a game does not affect the game in which a network cheats; it affects subsequent games. Thus, a state of history of a player's behavior in similar interactions must be kept.

We define a pricing mechanism consisting of variable $\beta_i \in [0,1]$, which represents the probability of being truthful, and it may adaptively modify the payoffs of a player. The value of β_i is adjusted at the end of a network-network interaction, according to the player's behavior, i.e., whether the network declared its real costs or whether the network lied[6], using a *punishment factor* $\gamma \in [0,1]$ set by the converged platform administrator[7].

Thus, based on $\beta_i^{previous}$, i.e., the previous value of β_i for network i, is defined to be:

$$\beta_i = \begin{cases} \beta_i^{previous} - (\beta_i^{previous} \cdot \gamma), & \text{if network } i \text{ is caught lying} \\ \beta_i^{previous} - (\beta_i^{previous} \cdot \gamma) + \gamma, & \text{if network } i \text{ is truthful.} \end{cases} \qquad (3.10)$$

Equation 3.10 defines β_i such that on the one hand it decreases fast when cheating behavior is observed, and on the other hand increases slowly when network i is truthful, aiming to motivate the players of the bargaining game to remain truthful since the less frequently a player cheats, the closer to 1 its β_i is. The value of $\gamma = \frac{1}{10}$ is reasonable since it allows the faster decreasing and slower increasing behavior of β_i. Next, we plot the general behavior of β_i when $\gamma = \frac{1}{10}$. Thus, Figure 3.1 illustrates the general form of β_i as it increases from 0 to 1 and decreases back to 0. Given β_i, the administrator sets the payoff of

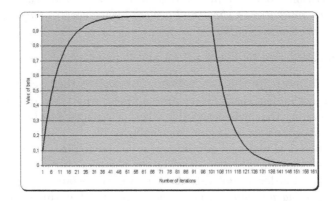

Figure 3.1: General form of β_i (*increasing and decreasing*).

---

[6]We consider that a revelation of the real costs of the two access networks is always possible at a later stage of the procedure (e.g., after session termination).

[7]In our simulations (Section 3.4.4) we set $\gamma = \frac{1}{10}$ as a tradeoff between harsh (e.g., $\frac{1}{2}$) and light (e.g., $\frac{1}{100}$).

network i to be:

$$\pi_i = \frac{1}{2}(\Pi + \beta_i \cdot c_i - \beta_j \cdot c_j), \qquad (3.11)$$

where $i, j \in [2], i \neq j$. The players are motivated to declare their real costs, since any cheating, which the platform administrator is able to detect following the play of the payment-partition game, would decrease β_i, affecting any future payoffs from such procedure. An evaluation of the Bayesian form of the *payment-partition* game including β_i in the players' payoffs is given in the next section.

Remark 9. *The actual one-shot payment partition is in practice simply a configuration imposed on the two interacting networks. In the Bayesian form of the game, each network has a memory of its own behavior during past participations in interactions with other networks in payment-partition games. Thus each network keeps its own internal state information for its actions, but this does not constitute a history of the game, since the game actions are not directly affected by the internal state maintained by each player. Furthermore, the decision of the opponent player or the history of the game does not depend on this state directly. Indirectly, the action taken by each player at a given interaction, may be such as to reflect the information obtained from this internal state, but this is according to the strategy employed by each network.*

3.4.4 Evaluating the game

In this section we evaluate the Bayesian form of the *payment-partition* game between two networks cooperating to support a service session requesting service continuity guarantees. The payoffs of the game are based on the Nash Bargaining Solution, as this has been demonstrated in Section 3.4.2, and further includes the term β_i, described in the same section, in order to motivate the networks to cooperate. The payoff to network i, defined in Equation (3.11), is reproduced next:

$$\pi_i = \frac{1}{2}(\Pi + \beta_i \cdot c_i - \beta_j \cdot c_j).$$

We first run the game with the value for β_i always equal to 1, i.e., when no punishment is imposed for lying about costs. The strategies used for the evaluation of the game for Case 1 (no punishment imposed), involve three different strategies for each player, where both networks are allowed to be truthful or lie as follows:

Case 1 Strategies:

1. The player always declares the real costs.

2. The player always lies about its real costs.

3. The player randomly lies 50% of the time and is truthful the rest of the time.

Subsequently, we run another set of simulations but allow the value of β_i to vary, i.e., we allow punishment based on the history of the network's previous actions. Thus, for Case 2 (with punishment imposed), one more strategy is added to allow the user to monitor β_i and make a decision according to its value. This additional strategy is the following.

Case 2 Additional Strategy:

4. The player monitors its β_i and only lies if the value of β_i is high, in order to minimize the effects of cheating on its payoff.

The numerical values used for Π, c_i, c_j in case the networks are truthful or lying, obey the payoff relations given in Table 3.4 and in Table 3.5 and the overall model of the payment-partition game as described and resolved in Section 3.4.2.

Specifically, the payment to be partitioned is equal to 10, and the costs vary from 3 to 5; 3 and 4 for the two types of networks when they are truthful, 4 and 5 for the two types of networks when they are lying (they claim costs of one unit more than the actual cost is). Future work can include an investigation of how the obtained results would be affected by different cost numbers, e.g., larger costs and unequal lying amounts.

The payoffs vary based on the type of each network as well as on whether the network has declared its real costs or has cheated. The network types are: $1 = lower\text{-}cost$ or $2 = higher\text{-}cost$, as these are explained in Section 3.4.3. The four strategies that each network may employ in a simulation run are denoted as follows: (1) always declare its real costs - indicated as *Only D*, (2) always cheat - indicated as *Only C*, (3) randomly declare real costs or cheat with a probablity of 0.5 for each option - indicated as *50% C/D*, and (4) only cheat if the value of β_i is high, i.e., cheat only if $\beta_i \geq 0.9$ else declare real costs - indicated as *Cheat-if-beta-high*.

First, we present the results for Case 1, when the value of β_i is equal to 1, i.e., when there is not punishment for lying. Table 3.8 and Table 3.9 for player 1 and player 2 respectively, show the cumulative payoffs from the payment partitions that the specific player receives for each strategy combination over 100 simulation runs.

Table 3.8: Payment partition payoffs for player 1, $\beta_1 = 1$

Network 2 Strategies	Only D	Only C	50% C/D
Network 1 Strategies			
Only D	5079.04	4578.54	4829.91
Only C	5579.04	5078.89	5329.42
50% C/D	5329.82	4829.30	5077.09

Considering the rows for Player 1, row 2 in Table 3.8 is the highest, showing that the best strategy for Player 1 is *Only C*. Also, for Player 2, we consider

Table 3.9: Payment partition payoffs for player 2, $\beta_2 = 1$

Network 2 Strategies	Only D	Only C	50% C/D
Network 1 Strategies			
Only D	4920.96	5421.46	5170.09
Only C	4420.96	4921.12	4670.59
50% C/D	4670.18	5170.7	4922.91

the columns in Table 3.9. The highest is column 2, thus for Player 2 the best strategy is also *Only C*. We observe from these results, that the highest payoffs are achieved for each player when he lies and does not declare his real costs, while at the same time his opponent declares the real costs. In any case, for each strategy of the opponent a player's best response is always to lie. Therefore, these results reinforce the theoretical findings of this model, i.e., that if no punishment mechanism is adopted for handling cases when players lie about their real costs, then the players are not motivated to be truthful and declare their real costs.

Next, we present the results for Case 2, when the value of β_i is allowed to vary according to player i's actions. Each player's initial value of β_i in the simulations is randomly generated and then adapted to the player's play according to the selected strategy. This initial value is indicated in the results for reasons of completion. Overall, we have run 100 simulations with different initial values for β_i and with a randomly generated number of iterations. The number of iterations for each simulation run is indicated in the results as R.

Table 3.10 illustrates some simulation statistics that show how many times in the simulation each network is of either type *lower-cost* or type *higher-cost* respectively (network types defined above), indicating that there is a uniformity in the generation of the player's types, giving us confidence in the fact that the two networks behave similarly and by examining one of them, we may draw similar conclusions for the other.

Table 3.10: Statistics for networks 1 and 2 regarding generation of types

Network 1 Type	Network 2 Type	Number of Times
lower-cost	*lower-cost*	27
higher-cost	*higher-cost*	26
lower-cost	*higher-cost*	21
higher-cost	*lower-cost*	26

Firstly, we show how the values of β_i evolve with time for network 1 (since corresponding values for network 2 are similar), for each of the four strategies that a network may employ. For each network i we keep its strategy fixed for each of the four plots but we allow the opponent's strategy to vary, plotting eventually the evolution of β_1. Figure 3.2 shows the variation of β_1 for each of the four employed strategies, indicated by a different line.

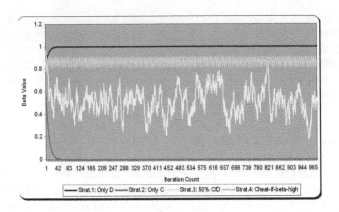

Figure 3.2: The variation of β_1 (first 1000 measurements).

We observe that when Player 1 employs the *Only D* strategy, its β_1 achieves the values equal to 1 since the network's strategy is to always declare its real costs, causing the values of β_1 to climb up to 1 and remain there until the end of the simulation. When network i employs the *Cheat-if-beta-high* strategy, the values are also quite high, oscillating between 0.8 and 0.9 because the network only cheats if β_1 is high, and since cheating causes β_1 to immediately decrease, the network declares its real costs, the value of β_1 increases again and when it becomes high enough to allow for cheating, network 1 cheats again and so on, resulting in an oscillating behavior for β_1. The values for the *50% C/D* strategy are more random and vary around 0.5 ranging from around 0.2 to around 0.8, since half the time network 1 cheats and half the times it declares its real costs and this is decided in a random manner. Finally, the *Only C* strategy where we observe that the values of β_1 converge to 0. The reason for this is that the continuous cheating behavior of network 1 causes β_1 to be decreased by the administrator after every period.

Tables 3.11, 3.12 present the averages of the cumulative payoffs from each simulation run, for all 100 simulation runs. The results are presented for the two networks (network 1 and network 2) partitioning the service payment. Tables 3.11, 3.12 illustrate through the similarity in the related payoffs that the behavior of the two networks is similar. This is due to the fact that the two networks are motivated by the same incentives and use the same set of strategies. In addition, the generation of network types in the simulation runs is very similar for the two players, implying that on average the payoffs are eventually very similar for the two networks.

We observe that in all simulation runs, the second strategy, i.e., always to cheat is the least profitable strategy for both networks, regardless of their types and initial values of β_i. This is because of the presence of β_i in the

Table 3.11: Average network 1 payoffs from payment-partition game

Avg. $R = 311.67$	Min. $R = 4$	Max. $R = 1749$			
	Network 2 Strategies	Only D	Only C	50% C/D	Cheat-if-beta-high
Network 1 Strategies					
Only D		1570.83	1855.16	1879.71	1591.25
Only C		1265.66	1558.72	1405.64	1301.49
50% C/D		1454.89	1718.82	1566.51	1479.84
Cheat-if-beta-high		1548.79	1827.05	1653.01	1569.23

Table 3.12: Average network 2 payoffs from payment-partition game

Avg. $R = 311.67$	Min. $R = 4$	Max. $R = 1749$			
	Network 2 Strategies	Only D	Only C	50% C/D	Cheat-if-beta-high
Network 1 Strategies					
Only D		1545.84	1251.89	1436.95	1525.41
Only C		1851.01	1557.94	1711.02	1815.18
50% C/D		1661.77	1397.84	1550.16	1638.83
Cheat-if-beta-high		1567.87	1289.62	1463.66	1547.44

payoffs, which punishes the choice of cheating, by detecting such an action after any iteration. On the other hand, the other three strategies, which include actions of declaring the real costs, i.e., being truthful are more profitable strategies. Specifically, the first strategy of always being truthful is the most profitable strategy illustrating how β_i rewards truthfulness, motivating the player to follow strategy *Only D*, i.e., always declaring the real costs. In addition, we observe that the fourth strategy of cheating only if β_i is high, generates comparable payoffs to the *Only D* strategy. This shows that even if the network decides to employ a strategy, which will allow the network to cheat a few times much less than the times it cheats when employing the 50% C/D strategy, the best option in terms of payoffs is still to be always truthful.

Furthermore, it is interesting to note that for each network, while it is more profitable to be truthful, i.e., to declare the real costs, the highest payoffs are accumulated when the opponent decides to cheat, while a network is truthful. Therefore, if a network believes that the opponent is more likely to cheat, it is very profitable to always be truthful, i.e., to only declare real costs.

In order to get a better sense of each partition, we provide for each combination of strategies, the percentage partition of the total payment given to the two networks, considering only the amount given and not the amount that would be given if they had cooperated (Table 3.13). We observe that the partitions, when the same strategies are used by the two networks, are equal or very close to 50% of the amount given, whereas, for the rest of the combinations using the strategy of always declaring the real costs, results in the greatest partition as it also seen by the corresponding payoffs.

Table 3.13: Payment partitions for each strategy combination

Network 1 Strategies	Network 2 Strategies	Only D	Only C	50% C/D	Cheat-if-beta-high
Only D		50.4%,49.6%	59.7%,40.3%	56.7%,43.3%	51.1%,48.9%
Only C		40.6%,59.4%	50%,50%	45.1%,54.9%	41.8%,58.2%
50% C/D		46.7%,53.3%	55.1%,44.9%	50.2%,49.8%	47.5%,52.5%
Cheat-if-beta-high		49.7%,50.3%	58.6%,41.4%	53%,47%	50.3%,49.7%

In conclusion, using β_i as a pricing mechanism, motivates the interacting networks to declare their real costs, so as to achieve the highest possible payoffs from their interactions.

References

[1] John F. Nash, "The Bargaining Problem," Econometrica vol. 18, no.2, pp. 155-162, April 1950.

[2] John F. Nash, "Two-person cooperative games," Econometrica vol. 21, no. 1, pp. 128-140, January 1953.

[3] Abhinay Muthoo, "Bargaining Theory with Applications," Cambridge University Press, Cambridge, UK, 2002.

[4] Ariel Rubinstein, "Perfect Equilibrium in a Bargaining Model," Econometrica vol. 98, no. 1, pp. 97-109, 1982.

[5] Noam Nisan, Amir Ronen, "Algorithmic Mechanism Design," Games and Economic Behavior vol. 35, no. 1-2, pp.166-196, April 2001.

[6] Roger B. Myerson, Game Theory: Analysis of Conflict, Harvard University Press, Cambridge, Massachusetts, 2004.

[7] Herbert Gintis, Game Theory Evolving: A Problem-Centered Introduction to Modeling Strategic Interaction, Princeton University Press, Princeton, New Jersey, 2000.

[8] Avinash Dixit, Susan Skeath, Games of Strategy, W.W. Norton & Company, New York, 1999.

[9] Peter Cramton, Yoav Shoham, Richard Steinberg, eds., "Combinatorial Auctions: Iterative Combinatorial Auctions," chapter by David C. Parkes, MIT Press, 2006.

[10] W. E. Walsh, M. P. Wellman, "A Market Protocol for Decentralized Task Allocation: Extended Version," Proceedings of the Third International Conference on Multi-Agent Systems (ICMAS-98), 1998.

[11] W. E. Walsh, M. P. Wellman, P. R. Wurman, J. K. MacKie-Mason, "Some economics of market-based distributed scheduling," Proceedings of The Eighteenth International Conference on Distributed Computing Systems (ICDCS-98), Amsterdam, The Netherlands, 1998.

[12] Rajdeep Dash, Nicholas Jennings, David Parkes, "Computational-Mechanism Design: A Call to Arms," IEEE Intelligent Systems, Special Issue on Agents and Markets, pp. 40-47, November 2003.

[13] David Parkes, "Classic Mechanism Design. Iterative Combinatorial Auctions: Achieving Economic and Computational Efficiency (Chapter 2)," University of Pennsylvania, May 2001.

[14] Theodore Groves, "Incentives in Teams," Econometrica, vol. 41, no. 4, pp. 617-631, July 1973.

[15] Martin J. Osborne and Ariel Rubinstein, "A Course in Game Theory," Massachussetts Institute of Technology, Massachussetts, USA, 1994.

[16] J. C. Harsanyi, "Games with Incomplete Information Played By Bayesian Players, Parts I, II, and III," Behavioral Science vol. 14, pp. 159-182, 320-334, 486-502, 1967.

Chapter 4

Cooperation for many: Spatial Prisoner's Dilemma and games in neighborhoods

4.1 Introduction

In this chapter we consider how cooperation can evolve in a *neighborhood* of players, i.e., a group of spatially proximal players. In such situations we deal with the interaction of each player with the rest of its *neighbors*. We revisit the Prisoner's Dilemma game to investigate the evolution of cooperation in such a spatially interactive situation. The original Prisoner's Dilemma game consists of the interaction between two players as we have seen in Chapter 2, where cooperation may evolve despite their selfish nature. Since in this case we are dealing with a neighborhood, i.e., a group of selfish players, where each player can interact with all of its direct neighbors, we focus on a variation of the classic Prisoner's Dilemma known as the *spatial* Prisoner's Dilemma [8].

Spatial game models take geometry into effect by locating the players that topologically form a two-dimensional lattice of usually nine players arranged in three rows and three columns. This is important because in spatial games, interactions take place only with immediate neighbors, and then go on to interact further with adjacent neighborhoods, i.e., groups of individual players that include only direct neighbors. In a spatial Prisoner's Dilemma game, we investigate the spatial coexistence of cooperating and defecting players, as their behavior affects the evolution of the neighborhood itself, where a few cooperative neighbors can invade the rest of the population modelled and induce cooperative strategies, to previously defecting adjacent neighborhoods.

We look for those behaviors that may result in equilibrium states and since we are talking about a group it is important that these solutions are also socially efficient, i.e., they benefit the group itself as well as the individual players. In the proposed scenario we will not only make use of the spatial version of the Prisoner's Dilemma but also of group strategies for Prisoner's Dilemma type of games. In the following sections we provide an overview of the spatial Prisoner's Dilemma game and of what we mean by group strategies, so that we can combine these two elements into a solution for the proposed illustrative scenario, which deals with reducing interference for wireless deployments in dense urban environments.

4.2   Spatial version of the Prisoner's Dilemma game

The spatial Prisoner's Dilemma is the spatial variant of the iterated Prisoner's Dilemma game [8]. It is a simple yet powerful model for the problem of cooperation versus defection in groups where each member of the group makes its own decision on the actions to take and is individually rewarded according to these decisions. This is unlike the coalitional types of game that we present in Chapter 5, which are also games of groups, but the payoffs are assigned to groups and not to individual players. In the spatial Prisoner's Dilemma game, each player's actions affect the payoffs for the rest of the players in the group as well as its own payoffs, since the payoff function of each player includes variables that are affected by the decisions taken by the other group players. As in the two-player non-spatial version of the Prisoner's Dilemma game, the spatial Prisoner's Dilemma game also demonstrates how altruism can become exploitation for personal gain, since although cooperation seems to be an option toward the social and not individual best payoff, the analysis shows that in the long run, cooperation is the best option in terms of individual payoff as well. The idea behind the spatial version of the game is to investigate how cooperation can evolve in an interacting population of topologically proximal individuals learning from each other by interactive experience.

Initially the population consists of a mix of cooperators and defectors interacting according to their topological placement. After each round of the game, the payoff for each player is determined based on its strategy and the payoff function, which is affected by the strategies of the rest of the players. For each subsequent round a player determines its new strategy by selecting the most favorable strategy for itself when interacting with its neighbors. Each player only locally interacts with its neighbors. According to each particular scenario and employed strategies, the spatial coexistence of cooperators and defectors is investigated and equilibria for the neighborhood are deduced, by calculating the averaged payoffs of each player when playing against its neighbors iteratively.

The original spatial Prisoner's Dilemma, known as the evolutionary Prisoner's Dilemma was introduced by Axelrod [8] to study the emergence of cooperation rather than exploitation among selfish individuals, where the players with the highest payoffs in a round are replicated in the following round. The new players in a single round are collectively referred to as a *generation*. Simultaneously, the *older generation* players are allowed to disappear. Thus the game demonstrates the evolution of a *species* of players, since only the strongest strategies survive, i.e., the strategies that result in the highest payoffs where average values are initialized at the end of each generation.

The spatial Prisoner's Dilemma game is played on a square lattice by monitoring when the players are interacting with their neighbors and which strategy they choose in each interaction, which results in appropriate payoffs. As expected a player can follow one of the two strategies: to cooperate or to defect. In the spatial version however, the players are updated in random sequence and have a chance to adopt one of the neighboring strategies with a probability depending on the payoff difference. In the simplest form of this game, the players located on a square lattice can follow only two simple strategies, either to always cooperate or to always defect, unlike the more complicated strategies we have seen in the iterated version of the Prisoner's Dilemma game, like Tit-for-Tat for instance. Each player plays the Prisoner's Dilemma game with its neighbors, where the neighbors may be defined in two different ways. In the first, only the direct neighbors are taken into account, i.e., players who are topologicaly adjacent, whereas in the second, the neighborhood includes the direct neighbors as well as the neighbors from directly adjacent neighborhoods (in which neighborhoods the players are topologically adjacent).

4.3 Group strategies for the Prisoner's Dilemma game

It is often useful to model games in groups in order to investigate the effects on the behavior of the whole group when interactions between the group members affect the outcomes of the individual group players. Modelling Prisoner's Dilemma as a group game considers the interactions between any two players in the group as we form groups such that any player's decisions are affected by the decisions of all the group members. Therefore, it is possible for a player to identify group members and consider playing strategies in cooperation with these group members to increase its own payoff, and as a consequence the payoff of the group.

The strategies involve a series of discrete moves by the players with each move leading to a change for the group in the game. Group strategies are based on the power of the group to gain payoff against opponents of the group, thus they work better for groups of a larger number of players and have poorer performance for groups of fewer players. A Prisoner's Dilemma game with many players, where any two players may interact according to the payoff matrix of cooperation against defection is ideal for group strategies, which are cooperative in nature, since the game promotes cooperation in its iterated form.

The payoffs that result from group strategies to the group players are not different from the traditional values in a prisoner's dilemma game, in the way that the payoff is obtained from a single interaction with one opponent. There exists a payoff A, for defecting when the opponent is cooperating, a payoff D, when cooperating but the opponent is defecting, a payoff C when both players defect, and a payoff B when both players cooperate with each other. The actual numerical values may vary from experiment to experiment but their relationship always obeys the rules for the payoffs of Prisoner's Dilemma, i.e., the inequalities $A > B > C > D$ and $2B > A + D$ are always kept. The second one ensures that cooperating twice pays more than alternating one's own defection with allowing oneself to cooperate while the opponent is defecting, i.e., alternatively exploit and be exploited.

Different numerical values are often used in testing the iterated Prisoner's Dilemma game that keep the above relationships. An example set of values could be a set where $D = 0$, $C = 1$, $B = 3$, *and* $A = 5$, a set used in some of the iterated Prisoner's Dilemma tournaments. Iterated Prisoner's Dilemma tournaments offer the opportunity to researchers that work with the Prisoner's Dilemma as a scientific tool, to experience the interaction of players employing specific strategies against players that may employ any strategy, to show through the cumulative payoffs the strengths of different strategies in terms of their outcomes in such environments. The empirical results of the iterated Prisoner's Dilemma tournaments organized by Axelrod have influenced the application of game theory in many scientific areas, and in this book we explore such outcomes in situations found in the networking field. It has been shown that adaptive players, learning from the games in which they are involved, are more likely to survive than non-adaptive players in evolutionary iterated Prisoner's Dilemma games [9], and we have shown an example of an adaptive strategy in Chapter 2.

Historically, the first tournament consisted of 14 entries that interacted with each other in a round robin iterated Prisoner's Dilemma game. Randomly choosing whether to cooperate or defect was also used as a strategy. Every entry interacted with every other entry, and the procedure was repeated five times in order to smooth out the effects of the random strategy. The strategy that won this first tournament was the strategy known as *Tit-for-Tat*, a strategy we have already studied in Chapter 2, where the player employing the strategy starts by cooperating, and then acts in whatever way its opponent acted on the previous iteration. Tit-for-Tat was the winning strategy in the second tournament as well, where 62 entries competed and the random strategy was once more participating in the group.

After the second tournament, the ecological tournament took place, where the spatial Prisoner's Dilemma (a.k.a., evolutionary Prisoner's Dilemma) was played and the entries were the same as in the second tournament. The difference from the second tournament was that in this evolutionary version of the game, players would adopt the most successful strategy of their neighbors at the end of a round robin set of interactions, which was referred to as a *generation*. The tournament consisted of 1,000 generations and the winner was again Tit-for-Tat.

From these three first tournaments some elements stood out in terms of what characterizes a good strategy. The strategy should employ *niceness*, i.e., never be the first to defect. Another element is that of *provocability*, i.e., get mad quickly at defectors and retaliate. However, besides being provocative, a strategy should also be *forgiving*, i.e., not hold a grudge once the anger has been vented. Finally, the element of *clarity*, i.e., act in ways that are straightforward for others to understand, also enhances a given strategy for success in such tournaments. These are all elements that are found in the winning Tit-for-Tat strategy.

The reasoning behind these successful strategy elements is the following: in any iterated Prisoner's Dilemma game with a finite number of iterations (the number of which is known to all players) the only Nash Equilibrium is mutual defection, thus player 1's defecting in all iterations is the best response to player 2's defecting in all iterations, but it is not the dominant strategy. This implies that even though the action sequence of defecting in all iterations is the best response to the opponent's action sequence of defecting in all iterations, it is not the best response to all possible action sequences (strategies). Examples of such strategies include: *random* strategy where in each iteration an unbiased coin is flipped to decide between cooperation and defection, *tit-for-two-tats* strategy, which start with cooperation in the first two iterations and defects only if the opponent defects in the previous two iterations, or *Grim Trigger* strategy, where the player cooperates but if the opponent defects, the player defects forever. Given the elements identified to best response to such strategies, i.e., strategies that combine cooperation and defection actions during the same game, in the subsequent tournaments several competitors submitted group strategies, i.e., strategies that use the members of the group to increase the payoff for the whole group.

Group strategies became popular in the 2004 and 2005 IPD tournaments [10], because they performed extremely well and defeated well-known strategies in round-robin competitions. These mechanisms identified an opponent according to its response to a certain sequence of cooperate and defect actions. In particular, the authors in [11] show that the members of such a group strategy are able to use a pre-arranged sequence of moves that they make at the start of each interaction in order to recognize one another, and that by coordinating their actions they can increase the chances that one of the team members wins the round-robin style tournament. The Southampton strategy was the most successful group strategy managing to win a tournament even against the many-times winner Tit-for-Tat strategy.

Particularly in the 2004 tournament, a team from the University of Southampton introduced a group of strategies, which were designed to recognize each other through a known series of 5 to 10 moves at the start of each game. Once 2 Southampton players recognized each other, they would act as a "master" or a "slave." A master will always defect while a slave will always cooperate in order for the master to win the maximum payoff. If a Southampton player recognized that the opponent was not a partner, it would immediately defect to minimize the score of the opponents. The Southampton group strategy is modified and used in this chapter to show how to motivate cooperation between group members in a neighborhood of a dense urban wireless deployment.

Another group strategy, which performed particularly well, is the CosaNostra strategy [12]. This strategy is based on the concept of one strategy exploiting another strategy, to achieve a higher total score in an iterated Prisoner's Dilemma tournament scenario. The idea is to deliberately extract cooperative moves from a strategy while playing defect, which results a higher payoff for the exploiting strategy. Since opponents would avoid to keep cooperating with a defecting opponent, eliciting cooperation while defecting within the group is desirable as it increases the group payoff, in particular for the leader of the group. Thus, a large number of group members that are willing to cooperate while their leader defects gives a high advantage to the leader in a tournament. In fact, the group members that are not leaders should be able to recognize their leader and not let any other opponent exploit them.

Therefore, the critical part of CosaNostra group strategy is the identification of opponents, the way in which leader detects a group member and a group member detects a leader. The strategy employs sequences of defections and cooperations, which both sides use to mutually establish, and check, identities. If the leader is aware that he is not facing a group member, he must switch to a good non-group strategy like Tit-for-Tat, and if a group member is aware that it is not facing the leader, he must switch to the always defecting strategy. The weakness of this strategy is for either the leader or the group member to wrongly identify an opponent as their strategic counterpart and grant it an advantage or depend on predefined behavior, lowering their score in the process. Specifically, if the leader thinks it is exploiting a group member, it defects, but the opponent also defects, so the leader gets the second lowest possible payoff for the interaction. To address this, the leader should be able to recognize a defect and conclude that the opponent is not a group member, and switch to the appropriate strategy. If a group member thinks he is being exploited by the leader and cooperates, it is harder to identify that the player who is exploiting him is not the leader. CosaNostra solves the problem by varying intervals of identification moves, with the number of iterations between exchanges in one interval being communicated within the identification moves.

Several group strategies emerged that were based on the same principles as the Southampton and CosaNostra group strategies, with one player playing the role of the leader and his group members submitting to yield to the leader the greatest payoff. One such example includes the *EmperorAndHisClones* group strategy.

Remark 10 (General Characteristics of Group Strategies). *The group strategies discussed above exhibit some characteristics that make them successful as identified in [12]. There are different types of group strategies, like strategies where all group members are equals and treat each other nicely by always cooperating, while they play different strategies like Tit-for-Tat or Grim Trigger when they interact with members outside the group. Also, similar group strategies exist such as those, which treat all group members as equals and treat each other nicely, however they continually defect against all other strategies. On the other hand, there are other strategies, the Southampton strategy included, which don't treat all members as equals. There is one special group member, which is allowed to take advantage of all other members of his group by playing defect while they cooperate with him. The rest of the group members cooperate with each other, and play strategies such as Tit-for-Tat or Grim Trigger with players outside the group, or in some cases their strategy is to always defect against all other strategies. Group strategies seem to perform arbitrarily better than individual strategies, and that, under equal group size, group strategies where members are not equal to the member who acts as a leader can achieve arbitrarily higher payoffs (for the leader) than groups where all members are equals.*

4.4 Spatial games and group strategies: Reducing interference in dense deployments of home wireless networks

Urban residential areas are becoming increasingly dense with more and more home networks being deployed in close proximity. This scenario considers a dense urban residential area where each house/unit has its own wireless access point (AP), deployed without any coordination with other such units. In this situation, it would be much better if neighboring APs, i.e., APs that are physically close to each other would form groups, where one member of the group would serve the terminals of all group members in addition to its own terminals, while the other access points of the group can be silent or even turned off, thus reducing interference and increasing overall Quality of Experience (QoE). The fact that participating units are deployed without any coordination makes the overall QoE vulnerable to the selfish behavior of each unit.

The scenario makes use of aspects of the spatial Prisoner's Dilemma game in order to motivate cooperation in an attempt to propose a protocol where each unit operates in an equilibrium of the modified spatial Prisoner's Dilemma game, which we will refer to as *cooperative-neighborhood game*. In addition, we make use of a modified version of a well-known group strategy for an iterated Prisoner's Dilemma game, namely the Southampton group strategy, in order to motivate neighborhing APs to enter and remain in cooperation. The game theoretic analysis shows that using a game theoretic model to study the interactions, achieves the above goal and that the formation of such cooperative neighborhoods decreases interference for the participators due to the voluntary cooperation of the neighbors.

4.4.1   Scenario overview

The scenario considers a dense urban residential area where each house/unit has its own wireless access point (AP), deployed without any coordination with other such units. Lacking any control regarding the efficient utilization of the communication channel, it is quite common for a terminal served by one of the APs to be within the signal range of multiple alternative APs. Since all APs are in competition for the same communication resource (radio channel), and the current standards dictate that at any given time every terminal must be rigidly associated with one particular AP, this situation results in increased interference and consequently a low utilization efficiency of the radio resource.

In a dense deployment, it would be much better for individual APs that are in physical proximity to each other to form groups, where one member of the group would serve the terminals of all group members in addition to its own terminals, so that the other access points of the group can be silent or even turned off, thereby reducing interference and increasing overall Quality of Experience (QoE). These groups would include only members whose signal strength is sufficient to serve all group members, so that the access point that would be responsible of serving the terminals of a particular group or *neighborhood* could change on a rotating basis, to allow all group members to equally serve and be served. Since there is no centralized entity that can control the APs and force them to form cooperative groups, the creation of such groups must be able to arise from a distributed process where each AP makes its own decisions independently and rationally for the benefit of itself and its terminals. *Game theory* is an appropriate tool to model such decentralized schemes.

The scenario models the idea of cooperative neighborhhoods as a game and shows that a group cooperative strategy in equilibrium, i.e., a strategy for units to voluntarily participate in a group where members serve terminals on a rotating basis, has the property that a unit participating in the group strategy is more likely to gain more in terms of QoE, than a unit defecting from such cooperation. In fact, to maintain robustness against uncoordinated deployments in dense residential areas, we propose a protocol with point of operation the game theoretic equilibria of a game, which we will henceforth refer to as the *cooperative-neighborhood game*. The game theoretic analysis results in proposing a protocol for the participating units to operate in the associated equilibrium point of the game to achieve the reduced interference. Numerical results for the cooperative-neighborhood game provide evidence both to the robustness of the proposed protocol, as well as to the improved overall QoE.

In the following sections we provide considerations of the environment, by discussing wireless deployment in urban environments, as well as the use of game theoretic approaches in distributed situations. In particular, we describe the scenario under study in game theoretic terms and parallelize it to an iterated Prisoner's Dilemma, proposing a group strategy in a given neighborhood that aims to motivate cooperation through the payoffs received by each unit individually. Possible payoffs are given through a numerical comparison of the proposed cooperative strategy with a defecting strategy, and based on the conclusions from this comparison, we describe a new protocol that is based on the equilibrium point of the cooperative neighborhood game, and can serve to enable this type of cooperative communication.

4.4.2 Wireless deployments in urban environments

The density of wireless networks in urban residential areas is on the rise with more and more home networks being deployed in quite close proximity, enabled by the low cost and easy deployment of off-the-shelf 802.11 hardware and other personal wireless technologies. It is not uncommon for a wireless station to be within range of dozens of access points [1], while the IEEE 802.11 standard only offers 3 non-overlapping channels. In this sense, urban areas are becoming similar to campus-like environments, however, in organizations and campuses experts can carefully control and manage interference of access points by planning the setup of the network in advance [2]. On the other hand, wireless networks in urban residential environments have the following two characteristics that make their deployment more challenging:

1. The network is unplanned, thus aspects of planning such as coverage and interference cannot be controlled. Deployments are mostly spontaneous, resulting in uneven density of deployment.

2. The network is unmanaged, lacking aspects such as efficient placement of access points, radio channel assignment, troubleshooting and adapting to network changes, for example, traffic load, as well as security issues.

The authors of [1] use the term chaotic deployments or chaotic networks to refer to such a collection of wireless networks which are unplanned and unmanaged. However, they do mention advantages of such chaotic networks, for instance, easily enabling new techniques to determine location [3] or providing near ubiquitous wireless connectivity [4]. The main disadvantage of these chaotic deployments is that interference can significantly affect end-user performance, while being hard to detect [5].

In this particular scenario, we consider a solution based on *virtualization* among the interfering APs, where APs serve each others' clients on a rotating basis. The security implications of allowing association of clients across APs from multiple owners have been addressed and resolved in [6]. What we focus on and the reason that we apply game theoretic aspects to this scenario is to demonstrate that there exists *incentive* toward cooperation, and with the use of the game models we describe a communication framework to ensure that the APs are indeed motivated to provide service to each others' clients.

We consider the interactions between the individual homes/units in dense urban deployments of wireless networks.

Describing and analyzing interactions between independent, *selfish* entities is a situation that makes a good candidate to be modeled using the theoretical framework of Game Theory. As can be seen throughout this book, Game Theory provides appropriate models and tools to handle multiple, interacting entities attempting to make a decision, and seeking a solution state that maximizes each entity's utility, i.e., each entity's *quantified satisfaction*. We concentrate on the Prisoner's Dilemma game model and in particular the spatial Prisoner's Dilemma game. The Prisoner's Dilemma game [4] has been a rich source of research scrutiny since the 1950s, with the publication of Axelrod's book in 1984 [8] becoming the main driver that boosted the concept to the attention of other areas outside of game theory, as a model for promoting cooperation, and as such we have considered it in Chapter 2 and here as well. Recall from Chapter 2 that the Prisoner's Dilemma is basically a model of a game, where two players must decide whether to cooperate with their opponent or whether to defect from cooperation.

4.4.3   Cooperative neighborhoods

Currently, dense residential deployments of home wireless networks consist of uncoordinated APs that serve their terminals individually. The APs do not form groups and share the communication channel, which is an unmanaged common resource, resulting in a low utilization efficiency due to the competition between the APs and the interference it causes. This interference can be reduced if the APs can form groups according to their location, such that any APs belonging to the same group can serve any terminal associated with any of the other group members. It is possible for an AP to recognize its *neighborhood* from the signals it receives, having a knowledge of the required signal strength thresholds that would serve its terminals in a satisfying manner.

In such a neighborhood only one of the APs needs to assume the role of a leader, while the others can remain silent, and thereby minimize the interference and improve the overall quality of experience for all terminals involved. The role of the leader can be assumed on a rotating basis. Of course, in order to take part in a cooperative neighborhood, the APs need to be motivated to act cooperatively, i.e., have an incentive to be silent or turned off while it is the turn of another AP to serve, and to serve everyone's terminals once their own turn comes. We show how such a distributed protocol can be sustained in the equilibrium point of a game model where each AP can make an independent decision whether or not to participate in such a neighborhood.

The interactions in a cooperative neighborhood can be modelled as a game between the participating units, where each member of the group has two choices at any given time: (a) to cooperate with its group members or (b) to defect from cooperation.[1] Which of the two behaviors to select in each round depends on the strategy of behavior that a player has decided to follow during the repeated game. The strategy of each *player*, i.e., each unit, is selected such that it results in the highest possible payoff for the particular player. We refer to such interaction between any two neighbors as a *cooperative-neighborhood game*.

Consider a game between 2 neighboring units, interacting as follows.

Definition 4.4.1 (Cooperative-neighborhood game). *Let each unit have a choice between two actions, cooperate or defect, when interacting with a neighboring unit. Let cooperation have one of two outcomes: $\alpha = f(x')$, when the interacting unit also cooperates, and $\alpha' = f(x)$, when the interacting unit defects, where x represents the level of interference, $x > x'$, and α represents a quantification of QoE, $\alpha > \alpha'$. Let defection have one of two outcomes: $\beta = f(x')$, when the interacting unit cooperates, and $\beta' = f(x)$, when the interacting unit defects, where β, much like α represents a quantification of QoE, $\beta > \beta'$, and the relationship between α and β is expressed as follows. $\alpha' < \beta' < \alpha < \beta$. Given that both interacting units choose their actions, they both aim to maximize their payoffs.*

We draw a parallel between the cooperative-neighborhood game and a Prisoner's Dilemma type of game. The classical Prisoner's Dilemma decision is whether at any point in time to cooperate with an opponent or defect from cooperation. Both players make a decision without knowing the decision of their opponent, and only after the individual decisions are made, these are revealed. Mutual cooperation has a reward for both players, specifically the second highest payoff that can be achieved by a player. However, such decision entails the risk that in case the other player defects, then the cooperative player will receive the least possible payoff. Given the risk of cooperation, it is very tempting to defect because if the opponent cooperates, then defecting will result in the highest possible payoff, although, if the other opponent also defects then the payoff received by both players will only be slightly better than the worst possible payoff. In Definition 4.4.2 we remind the reader of the specific elements of the classic Prisoner's Dilemma kind of game.

Definition 4.4.2 (Prisoner's Dilemma type of game). *[4] Consider an one-shot strategic game with two players in which each player has two possible actions: to cooperate with his opponent or to defect from cooperation. Furthermore, assume that the two following additional restrictions on the payoffs are satisfied:*

---

[1]Note that units may be a part of more than one neighborhood, i.e., receive a good signal from peers that are in different neighborhoods. In the basic case that we look at in this section, for simplicity, we consider that each unit is part of only one neighborhood. In a subsequent section we will consider a larger population.

1. *The order of the payoffs is shown in Table 4.1 for each player $j \in \{1,2\}$ and is such that $A_j > B_j > C_j > D_j$.*

2. *The reward for mutual cooperation should be such that each player is not motivated to exploit his opponent or be exploited with the same probability, i.e., for each player it must hold that $B_j > \frac{A_j + D_j}{2}$.*

Table 4.1: General payoffs for the Prisoner's Dilemma

	Player 2 Cooperates	Player 2 Defects
Player 1 Cooperates	B_1, B_2	D_1, A_2
Player 1 Defects	A_1, D_2	C_1, C_2

Then, the game is said to be equivalent to a Prisoner's Dilemma *type of game.*

Next we provide a formal proof that the cooperative-neighborhood game is equivalent to a Prisoner's Dilemma type of game by making use of Table 4.2, which illustrates the mapping between a unit's payoff in a cooperative-neighborhood game and the payoffs in a Prisoner's Dilemma type of game.

Table 4.2: The mapping between the payoffs for either unit in the cooperative-neighborhood game and the payoffs for either player in the Prisoner's Dilemma type of game

	Unit Payoffs
A	$\beta = f(x')$
B	$\alpha = f(x')$
C	$\beta' = f(x)$
D	$\alpha' = f(x)$

Proposition 4.4.1. *The cooperative-neighborhood game (Definition 4.4.1) is equivalent to a Prisoner's Dilemma game.*

Proof. By Definition 4.4.1 we immediately conclude that:

Observation 2. *There are two possible actions for a unit: (i) to participate in a neighborhood in a cooperative way, and (ii) to defect from cooperation and not participate in the neighborhood.*

Observation 2 combined with Definition 4.4.2 implies that the actions of the players in the cooperative-neighborhood game match the actions of the players of a Prisoner's Dilemma type of game. In particular, Table 4.2 maps each unit's payoffs, to actions A, B, C, D, as defined in Definition 4.4.2. We proceed to prove:

Lemma 4.4.1. *Set A, B, C, D according to Table 4.2. Then it holds that $A > B > C > D$.*

Proof. By Definition 4.4.1 it is true that $\alpha' \leq \beta' < \alpha < \beta$. Since, β is mapped to A, α to B, β' to C and α' to D, then the relationship $A > B > C > D$ holds. □

We now proceed to prove that:

Lemma 4.4.2. *The cooperative-neighborhood game satisfies condition 2 of Definition 4.4.2.*

Proof. To prove the claim we must prove that the reward for cooperation is greater than the payoff for the described situation, i.e., for each player it must hold that $B > \frac{A+D}{2}$.

For each unit,

$$\alpha > \frac{\beta + \alpha'}{2}. \tag{4.1}$$

Since α and β are quantifications of an experience with little to no interference, then $\alpha - \beta$ is a very small value, as opposed to $\zeta - \alpha = \alpha'$, which is a larger value since the two values represent on the one hand the case of little to no interference and on the other hand the case of great interference in the experience. Therefore it must hold that:

$$2\alpha - \beta > \alpha'. \tag{4.2}$$

□

Observation 2, Lemma 4.4.1, and Lemma 4.4.2 together complete the proof of Proposition 4.4.1. □

The decision of what to do comes from the following reasoning: *If a player believes that his opponent will cooperate, then the best option is certainly to defect. If a player believes that his opponent will defect, then by cooperating he takes the risk of receiving the least payoff, thus the best option is again to defect.* Therefore, based on this reasoning, each player will defect because it is the best option no matter what the opponent chooses, i.e., to defect is the dominant strategy. However, this is not the best possible outcome of the game for both, i.e., mutual defection is not socially efficient. The socially efficient decision for both players would be to cooperate and receive the second best payoff.

The desirable cooperative behavior must be somehow motivated so that the players' selfish but rational reasoning results in the cooperative decision. In fact, even though mutual defection is the Nash Equilibrium solution for the one-shot game, because no previous or future interaction of the two players affects this decision, mutual cooperation may be motivated, not only as a socially efficient strategy but also as an equilibrium solution, from playing the game repeatedly, against the same opponent, if the opponents are not aware of the number of total iterations, i.e., if the believe that they are playing an infinite game. This is the idea behind the iterated Prisoner's Dilemma game, a repeated game model with an unknown or infinite number of repetitions[2].

The decisions at such games, which are taken at each repetition of the game are affected by past actions and future expectations, resulting in strategies that motivate cooperation. In fact, the model of a repeated game is designed to examine the logic of long-term interaction. It captures the idea that a player will take into account the effect of his current behavior on the other player's future behavior. The idea on which the basis for cooperation is built, is that if the game is played repeatedly, then the mutually desirable cooperative outcome is stable because any deviation will end the cooperation, resulting in a subsequent loss for the deviating players that outweighs the payoff from the finite horizon game (horizon of one or more periods).

Group strategy to motivate cooperation

In the scenario of a home wireless network deployment, we may consider the interactions to be infinite, i.e., we have no way of knowing if and when they will end. An additional aspect that needs to be considered is the location of each unit, which affects the group members it can select, i.e., the players in the same neighborhood that have strong enough signals to support all terminals handled by the neighborhood units. For simplicity, we first assume that each neighborhood ideally forms a separate group, i.e., all players in the neighborhood cooperate with each other and there are no partial coverage overlaps[3]. In section 4.4.3, however, we will further consider what happens when some of the neighborhood players do not participate in the group, i.e., defect from cooperation.

We henceforth consider group strategies against defectors, employing a *punishment* of a defector from a group, in order to strengthen motivation to remain in a cooperative group. It has been shown that group strategies can perform even better than individual strategies known to prevail in a wide range of cases, such as Tit-for-Tat [11]. We are inspired by the Southampton group strategy [11], where members of a group may assume one of two roles at any round of the game, and furthermore, employ methods to recognize group members, so that the strategy is *friendly* only toward their own group members. The Southampton strategy was the first strategy to win against Tit-for-Tat strategy in the Prisoner's Dilemma tournament in 2004 [10]. It is a group strategy, employing an identification mechanism that helps to distinguish group members from opponent players. Particularly, in every interaction

---

[2]The model of a repeated game has two kinds: the horizon may be *finite*, i.e., it is known in how many periods the game ends, or *infinite*, i.e., the number of game periods is unknown.

[3]In fact it has been shown that in a spatial Prisoner's Dilemma type of game, i.e., where many neighborhoods are considered, the players may be motivated again into a cooperative behavior [4].

with another player, the Southampton player performs a predetermined sequence of moves and uses the response of the competing player to decide whether it is a member or not. If the opponent is not also a Southampton player, the strategy of the Southampton player is to keep defecting.

In the Southampton strategy an interaction of two players where both employ the strategy leads to one of them defecting and the other cooperating to increase the payoff of the group. We propose a modified Southampton group strategy, where if both players are in the same group then they should both cooperate, since in our scenario the cost of a defection will have a negative impact to the group (i.e., increase interference). Two cooperating players, i.e., from the same group, may assume different roles in the case that one of them is the leader of the group and the other one is not, or have the same role if neither of them is the leader of the group. However, in both cases we propose that the two players cooperate with each other. The reason for this modification is that, unlike the Southampton group strategy, it is important that we ensure a socially efficient strategy, i.e., all the group members should have increased payoffs in the duration of the interaction and not simply the leader of the group in each interaction, i.e., by defecting while the rest of the members cooperate.

Definition 1.1.3 (Modified Southampton Group Strategy). *Two players employing this strategy and belonging to the same group, always cooperate with each other once they recognize each other as a group member, whereas interaction between a player employing this strategy and a non-member will cause the player employing this strategy to continuously defect.*

Evaluation of the Modified Southampton Group Strategy

This section briefly examines the numerical behavior of interacting players employing either the cooperative approach, i.e., participate in a group and employ the modified Southampton strategy, or the defecting approach and do not participate in the neighborhood's group, in fact defecting on each round. The evaluation is based on a MATLAB implementation of an iterated Prisoner's Dilemma type of game, where 9 players (the typical number for a single lattice used for a neighborhood in spatial Prisoner's Dilemma games) interact either as cooperators or defectors. Thus, the strategies are played against each other multiple times in order to evaluate the behavior of each strategy in terms of payoffs. The implementation of the game model was based on a publicly available MATLAB implementation of the Iterated Prisoner's Dilemma Game [3], which has been modified to support the proposed version of the Southampton strategy for the cooperative neighborhood game. Guidelines for the implementation of strategies for Prisoner's Dilemma type of games are outlined and explained in Chapter 6.

The implementation makes use of the following guidelines, set to reflect the payoffs of the interaction game as shown in Table 4.2. In each simulation run, 9 players play their strategies and get payoffs accordingly. The numerical payoffs are the following in a single round: if one player defects and the other cooperates, the first gets 4 and the other gets 0, if they both defect, each gets 2, and if they both cooperate each gets 3. Eight different sets of simulations are run with the first set of simulations considering that all 9 neighbors are members of the group and thus cooperators, the second set of simulations assumes one defector in the neighborhood, the third set assumes 2 defectors and so on until the 8 simulation set which assumes 7 defectors, i.e., the group only has 2 members in the neighborhood.

A randomly generated number of iterations was run repeatedly for each set of simulations to get cumulative payoffs for each combination of strategies playing against each other. Note that the strategies we have considered are the always-defect strategy for the defector and the modified Southampton strategy for the cooperator. The payoffs per strategy are eventually added and averaged to give the average cooperator and the average defector payoffs in each set of simulations, in order to investigate the effect on the payoffs of the group members as the number of defectors in a neghbourhood increase, given the 2 particular strategies are used. The reason we generate a random number of iterations in each simulation set and get the average, is to be able to include in our results behaviors that occur when the number of iterations is both small and large.

Figure 4.1 shows the average cooperator and defector payoffs in the 8 different simulations sets. We observe that cooperators enjoy overall higher

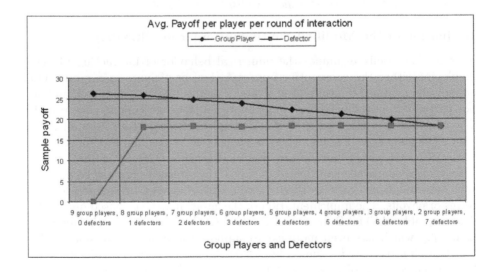

Figure 4.1: Numerical results of interacting cooperators and defectors in a neighborhood.

payoffs than defectors, however, as the defectors increase, the payoff of the group decreases and thus the payoff per cooperating player decreases as well. It is obviously more beneficial that no defector exists because the difference between a cooperative player's payoff when no defector exists and when at least one defector exists is significant. However, it is still more beneficial to cooperate, thus there exists motivation to form the coperative group by employing the *modified Southampton group strategy*. This behavior is observed because of the relationships between the payoffs, due to the equivalence to the Prisoner's Dilemma game, therefore even if changing the actual payoff values, the numerical motivation to form the cooperative group would still be apparent.

This makes sense since as defectors in a neighborhood increase, the interference increases and the quality of experience of terminals in units participating in the cooperating group, decreases. We may reach such conclusion by considering the relationship between the payoffs used in the simulation which matches the relationships between the payoffs in the cooperative neighborhood game.

Our current model is limited by the approximation that all players (access points) within a cooperative group are perfectly equivalent and interchangeable, and can serve each other's terminals with equal performance. The extension of the model and results to allow for more complex topologies with partial neighborhood relationships among the APs (i.e., where ranges of the individual APs overlap only partially), and their evaluation using a more realistic wireless networking simulation environment, are interesting extensions of this scenario.

4.4.4   A protocol for cooperative neighborhoods

The equilibrium point of the *cooperative neighborhood* game is that all members of a neighborhood cooperate by assuming either the role of the leader and serving all terminals of the neighborhood, or remaining silent to reduce interference while its terminals are being served by the leader. Therefore, based on the premise that there exists cooperation between members of a particular neighborhood, we next outline a distributed protocol by specifying the interactions between the neighborhood members in terms of message exchanges.

The protocol needs an identification phase so that an access point may know its neighborhood members. Let us primarily consider the case that such a cooperative neighborhood has not been set up yet. The first step is to broadcast a request to participate in the group. This is assumed to be a leadership move, i.e., the initiator to form a group becomes the first leader of the group (Figure 4.2). The cooperative members will respond to the request by communicating a list of access points which have strong signals to support their terminals and the leader will find the common set of these lists with its own list and broadcast the list of the final neighborhood members. After this phase each member of the group knows its neighbors. Now, asymmetric cryptogra-

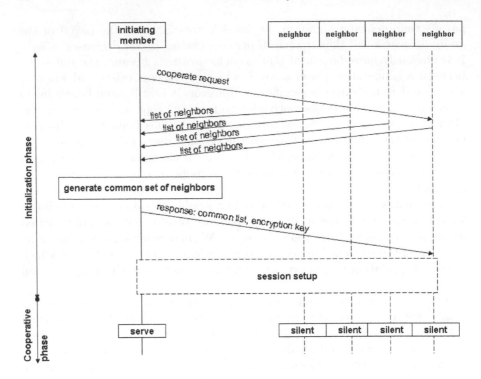

Figure 4.2: Initialization phase: requesting member becomes the leader.

phy can be used between the members, i.e., the leader may communicate an encryption key, to ensure that any further communication of session control information is private within the neighborhood.

In the case that the requesting member is not the initiator of a group then the leader will respond with a message identifying itself as a leader and including in the message the list of group members and the encryption key for the group (Figure 4.3). The rest of the neighborhood members will ignore the request if they are already participants of one of the other active cooperative neighborhoods. The requesting member needs to check whether it receives a strong signal from all group members and in case this is true, it responds with an OK to the leader and the session information is then communicated in a secure communication. The session is hence modified by the leader so that the new neighborhood member is added. Otherwise, in the case the new member cannot join the cooperative neighborhood, it does not respond to the leader and after a preset time interval elapses, the leader considers this as a *No* response.

Once the identification phase is completed and the session is set up or modified, each access point behaves in either of two ways: serves all terminals in a neighborhood or remains silent. The rotation phase of the protocol occurs in a timely manner according to the member number of each participant, by

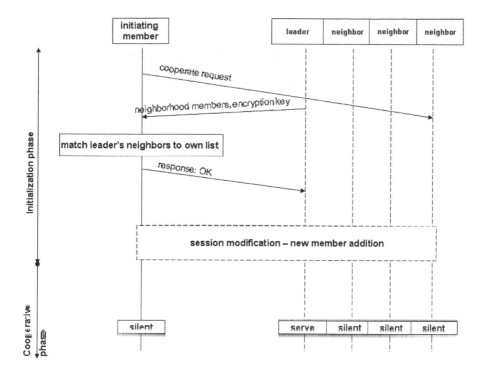

Figure 4.3: Initialization phase: requesting member joins an active neighborhood.

a session modification initiated by the leader. Figure 4.2 and Figure 4.3 show schematics of the setup phase of this protocol in the case that the requesting member becomes the leader and in the case the requesting member joins an active neighborhood.

4.5 Conclusions on neighborhood games

The chapter investigated the interactions between a group of players, as for example, wireless access points that operate in the same geographical region without any coordination. Using a game-theoretic model inspired by the iterated Prisoner's Dilemma type of game, it has been shown that the players are motivated to create alliances with their neighbors so as to serve their terminals jointly and in a coordinated manner, leading to an equilibrium where all neighbors act cooperatively. The numerical results show the value of the cooperation for the case of infinitely repeated interactions for topologically proximal units. Furthermore, to touch upon implementation aspects, we have outlined a protocol that can support the necessary coordination and deals with cooperation requests during the formation of a cooperative neighborhood.

References

[1] A. Akella, G. Jedd, S. Seshan and P. Steenkiste, *Self-Management in Chaotic Wireless Deployments*, In ACM MobiCom, pp. 185-199, 2005.

[2] A. Hills, *Large-Scale Wireless LAN Design*, IEEE Communications vol. 39, no. 11, pp. 98-104, November 2001.

[3] Intel Research Seattle, Place Lab, *A Privacy-Observant Location System*, http://placelab.org/, 2004.

[4] O. A. Dragoi and J. P. Black, *Enabling Chaotic Ubiquitous Computing*, Technical Report CS-2004-35, University of Waterloo, Canada, 2004.

[5] P. A. Frangoudis, D. I. Zografos and G. C. Polyzos, *Secure Interference Reporting for Dense Wi-Fi Deployments*, Fifth International Student Workshop on Emerging Networking Experiments and Technologies, pp. 37-38, 2009.

[6] J. Hassan, H. Sirisena and B. Landfeldt, *Trust-Based Fast Authentication for Multiowner Wireless Networks*, IEEE Transactions on Mobile Computing, vol. 7, no. 2, pp. 247-261, 2008.

[7] G. Kendall, X. Yao and S. Y. Chong, *The Iterated Prisoner's Dilemma: 20 Years On*, Advances in Natural Computation Book Series, vol. 4, World Scientific Publishing Co., 2009.

[8] R. M. Axelrod, *The Evolution of Cooperation*, BASIC Books, New York, USA, 1984.

[9] M. Nowak, A. Sasaki, C. Taylor and D. Fudenberg, *Emergence of cooperation and evolutionary stability in finite populations*, Letters to Nature, vol. 428, April 2004.

[10] W. M. Grossman, *New Tack Wins Prisoner's Dilemma*, http://www.wired.com/culture/lifestyle/news/2004/10/65317, October 2004.

[11] A. Rogers, R. K. Dash, S. D. Ramchurn, P. Vytelingum and N. R. Jennings, *Coordinating team players within a noisy Iterated Prisoner's Dilemma tournament*, Elsevier Theoretical Computer Science, vol. 377, pp. 243-259, 2007.

[12] W. Slany and W. Kienreich, *On Some Winning Strategies for the Iterated Prisoner's Dilemma.*

[13] G. Taylor, *Iterated Prisoner's Dilemma in MATLAB* , Archive for the "Game Theory" Category, http://maths.straylight.co.uk/archives/category/game-theory, March 2007.

Chapter 5

Cooperation for many: Payoffs to coalitions

5.1 Introduction

This chapter investigates the situation of cooperation between multiple selfish entities, but unlike Chapter 4, in this chapter we study the formation of coalitions, i.e., groups of entities that have a goal as a group and not as individuals; the goal being to increase the group's payoff. When situations arise that groups must be formed because payoff is paid to a group and not to an individual player, then the main strategical question is how to decide which group to join, or in the case groups are formed from scratch, how to form these groups. It is important to participate in a successful group in order to receive satisfying payoff, since there can be winning and losing groups in coalition games.

The real-life networking scenario that we illustrate is a situation, which involves multiple access networks cooperating to support a particular anticipated service demand for a certain period (e.g., supporting a multiparty multimedia service), and which none of the participating networks is prepared to handle on its own. For example, in the case that increased service demand exists, which none of the access networks is capable or willing to support on its own, this requires the participating access networks to cooperate in order to form appropriate coalitions that can serve the particular demand. In this case, an appropriate coalition formation game theoretic framework is developed to direct the selection and formation of the most appropriate coalition.

Toward this goal, we study coalition formation between the participating networks according to their available resources, given that the payment for a particular anticipated traffic demand is paid to the coalition, considering scenarios of both transferrable and non-transferrable payoff. Furthermore, we seek a fair payoff allocation rule for the networks participating in the winning coalition. We show that the coalition most likely to be formed may be identified and based on the identified coalition, the payoff allocation is fair. To achieve the corresponding payoff allocation, we propose a newly defined power index (a way to allocate weights to the players according to certain conditions) and compare it to existing power indices to show the improved performance for the particular cooperative situation. The idea of power indices will be discussed further in subsequent sections of the chapter.

This chapter focuses on the weaknesses and merits of coalitional game theory to model such a situation. The coalition formation process between multiple networks has been modelled as a new game that we refer to as the *Network Synthesis* game, in which individual networks with insufficient resources form coalitions in order to satisfy service demands. It is shown that the Network Synthesis Game is equivalent to the well-known *Weighted Voting Game*, which we describe informally in a subsequent section prior to its formal definition and equivalence proof to the *Network Synthesis* game.

We have previously mentioned that one of the most important aspects of the work on coalitional game theory is to figure out the best way to form a group or the best way to join a group, in order for the payoff that the group receives to be satisfying. Toward this goal, a comparative study of well-known *power indices* representing payoff schemes, is provided for the Network Synthesis Game. Furthermore, a new power index, the *Popularity Power Index* (PPI), is proposed based on conclusions from the study of well-known power indices, which associates the popularity of each network to the number of stable coalitions it participates in. The newly proposed power index, PPI, achieves fairness, in the sense that it only considers the possible coalitions that would be formed if payoffs were assigned proportionally to the networks' contributions.

An analysis of the coalition formation is provided for both transferable and non-transferable payoffs, in order to determine stable coalitions using the core and inner core concepts. It has been shown that from the well-known power indices we have studied, the most appropriate for the Network Synthesis Game is a power index that provides stability under the core concept, known as the *Holler-Packel Index (HPI)* [8]. Moreover, coalitions that would be formed using the newly proposed PPI to assign payoffs, are only coalitions that would be stable under the inner core concept, which is a widely used concept in coalitional games that we later define within the current chapter. Therefore, the PPI provides a simple payoff allocation method that is equivalent to a cooperative equilibrium solution of the Network Synthesis Game.

5.2   Games of coalitions

Coalitional games deal with the situation in which interactions occur between groups of players (coalitions), and thus actions are assigned to coalitions even though individual entities may consider their own preferences, especially when selecting a particular coalition in which to participate. Therefore, a coalitional model is characterized by its focus on what groups of players can achieve rather than on what individual players can achieve.

Furthermore, in coalitional games, the way a coalition operates internally, i.e., among its members, is not considered as important for a coalitional game, so that the outcome does not depend on such details. A solution concept for coalitional games assigns to each game a set of outcomes, capturing consequences for the participating coalitions. The solution defines a set of arrangements that are stable in some sense, i.e., that the outcomes are immune to deviations of any sort by groups of players.

In order to determine the solution to a coalitional game, we must first define the way payoffs are assigned to the various coalitions; such assignment can occur per group as a whole, or per group using a particular division arrangement within the group for its members. When payoffs are assigned per group, the players that participate in the same group are associated with the group's payoff and it is not defined how this payoff may be further partitioned among its members. This case of payoff assignment is referred to as transferrable payoff coalitional game. The alternative is known as non-transferrable payoff coalitional game, and in such model there exists a rule on how group payoffs are divided among participating players.

A well-known solution concept for a coalitional game model is the *core* concept. The core is a solution concept that requires that no set of players be able to break away and take a joint action that makes all of them better off. Overall, the idea of the *core* is analogous to that behind a Nash Equilibrium for a non-cooperative game, i.e., an outcome is stable if no deviation is profitable. In the case of the *core*, an outcome is stable if no coalition can deviate and obtain an outcome better off for all its members.

Following, we present the *Voting Game* model, one of the most well-known coalitional games with many applications, both in the scientific area of political science in its original form, but also in other scientific areas, modeling diverse scenarios. In our analysis it is used to model a real-life networking scenario. As the name implies, the original game presented dilemmas and decisions for a given voter on how to select the best candidate to cast a vote for.

5.3 The Voting Game

The Voting Game [2] is a game model that deals with election systems, i.e., ways in which representatives can be elected in a given election. The idea is inspired by the fact that different governments use different election systems, giving a better understanding into the idea behind such systems and offering insights on good strategies in the voting procedure. Therefore, the Voting Game is designed to provide a hands-on non-judgemental experience of various election systems. In this chapter we describe 5 different systems, that are often used as coalitional game models. In such Voting Games, the participants, i.e., the voters, are presented with their voting options, in this case for each of these 5 different election systems, and then they vote for the various candidates. The varying game models, representing the various real-life voting systems, give insight into how the election results differ under the different systems when everything else is the same. In the next few paragraphs, we describe the 5 different voting systems. One of these models, the Weighted Voting Game, is the model we later use in our real-life networking scenario.

First Past the Post Voting system   *First Past the Post* voting refers to an election won by the candidate(s) with the most votes. The winning candidate does not necessarily receive an absolute majority of all votes cast, i.e., does not receive more than half the votes, but it receives more votes than any of the other candidates. The first past the post voting method, also known as the "winner-take-all" system, in which the candidate with the most votes gets elected, can use either single or multiple selection ballots, therefore each voter can vote once or multiple times. A multiple selection ballot where more than one candidate can be voted for is also a form of first past the post voting in which voters are allowed to cast a vote for as many candidates as there are vacant positions. Again, the candidate(s) with the highest number of votes is elected. However, if the votes are for individuals, but the individuals belong to parties, then if the elections give *seats* to the winning candidates, then at the end the larger parties will gain a disproportionately large share of seats, while smaller parties are left with a disproportionately small share of seats. In such games, the coalitions would be defined among the voters, according to which candidate they vote for.

Majoritarian Voting system   The second system, known as the *Majoritarian Voting*, is the oldest electoral system, and also the simplest. This systems requires candidates to win an absolute majority (50% plus one vote) of votes to be elected. In simple words, to win a candidate is required to have simple numerical majority in an organized group of voters. In case none of the candidates receives this numerical majority, alternative mechanisms can be used to ensure that the winning candidate gets an overall majority of votes. Often the second ballot *majority-runoff* system is used, where candidates obtaining an absolute majority of votes (50% plus one vote) in the first round are declared elected. If this is not the case a second round is held between the two candidates who got the highest number of votes. Again, the coalitions are formed depending on which candidate a voter selects; however, the winning coalition in this case is only one that supports the candidate that manages to collect the majority of votes.

Proportional Representation system   *Proportional Representation* is when the percentage a political party gets from the popular vote, becomes the percentage of the seats that party receives in parliament. The candidates that actually get elected to sit in that parliament are usually decided by means of party lists, where the party lists its candidates in order, i.e., if it wins n seats, the first n candidates on its list are elected. Proportional Representation systems are fairer since the voters form coalitions to vote for parties and the winning party gains proportionally to the votes it received while in systems such as first past the post, the seats that a party receives is not proportional to the votes it receives since the voters support individual candidates.

Single Transferable Vote system The *Single Transferable Vote* is a voting system designed to achieve proportional representation through preferential voting. In this system, a vote is initially allocated to the voter's most preferred candidate, and then, after candidates representing the voters have been either elected or eliminated, any surplus of voters or unused votes are transferred according to the voter's stated preferences. The system minimizes *wasted* votes, provides approximately proportional representation, and enables votes to be explicitly cast for individual candidates rather than for closed party lists. Thus, coalitions of voters are formed according to preference and in support of particular candidates and not parties.

Weighted Voting system Weighted voting systems are voting systems based on the idea that not all voters are equal. Instead, it can be desirable to recognize differences by giving voters different amounts of *weights* concerning the outcome of an election. This is in contrast to the previously described election procedures, which assume that each member's vote carries equal weight. A weighted voting system needs to consider the players, i.e., the voters, the weights of the voters and the quota. A voter's weight w is the importance of his vote, i.e., it could be described as the number of votes the player controls when all votes are of equal weight. The quota q is the minimum number of votes required to succeed, e.g., in passing a motion. Any integer is a possible choice for the quota as long as it is more than 50% of the total number of votes. Each weighted voting system can be described using the generic form $[q : w1, w2, \ldots, wN]$, i.e., by specifying the game quota and the weights of each player for all players.

This chapter concentrates on this last voting system, the weighted voting, which will be used as the basis for the game model we refer to as Network Synthesis Game in our real-life scenario. It is important in such scenario to know and represent the power that each player has since these are not equal but weighted. The notion of power is often used for a player as a quantification of player's ability to influence decisions. For example, consider game $G = [6 : 5, 3, 2]$, where a motion can only be passed with the support of player $P1$, who has veto power. A player is said to have veto power if a motion cannot pass without the support of that player. On the other hand, a dummy is any player, regardless of his weight, who has no say in the outcome of the election.

A player without any say in the outcome is a player without power. Consider game $H = [8 : 4, 4, 2, 1]$. In this example, neither voter with weight 1 or 2 have any power. Neither can affect the passing of a motion, thus they are referred to as dummy players.

In each one of the cases described above, voting is basically a standard method of casting a preference in multi-agent environment, the agents in this case being the voters. Although, not obvious in its process, which is an individual process, voting allows the voters to make joint decisions by selecting the most suitable candidate from a given set. Thus, the voters form coalitions according to their preference. The motivation behind such a coalition formation process is not always the same, i.e., it could be to optimize the group utility or it could be to optimize their individual utility, in which case the voting process has to deal with manipulating the election rules for individual gain. Some voting systems are harder to manipulate than others, and in fact the different Voting Game models that represent different election systems are often classified according to the resistance of the voting rules to be manipulated for individual as opposed to social gain. While the possibility of manipulation by a single voter presents a grave concern from a theoretical perspective, as we have seen in some examples in the text already, in real-life elections this issue does not usually play a significant role: typically, the outcome of a popular vote is not close enough to be influenced by a single voter. However, a more significant problem that may be considered similarly to this is that of coalitional manipulation, where a group of voters coordinates their actions in order to affect the election outcome. We will consider the idea of power and Weighted Voting Games from a theoretical perspective in order to come up with a good model for the Network Synthesis Game used in our scenario, but we will not examine the coalition manipulation issue in our scenario.

5.4 Players' power to affect decisions in a coalition game

A player's weight is not always an accurate depiction of that player's power. Sometimes, a player with several votes can have little power. It depends mostly on the quota and the number of players. To clarify this, let's consider the example of game $I = [30 : 15, 15, 14]$. In this case, although player $P3$ has almost as many votes as the other players, his votes will never affect the outcome. Conversely, a player with just a few votes may hold quite a bit of power. Take for instance, game $J = [10 : 5, 4, 1]$. No motion can be passed without the unanimous support of all the players. Thus, player $P3$ holds just as much power as any of the other two players.

The formation of a coalition is greatly influenced by the players themselves in terms of how much they motivate such cooperations with other players. Power indices are often used as a coalition formation method in simple (0-1) games. A power index (or value) is commonly used to measure the influence of a player on the formation of coalitions and most importantly on the outcome of the game. The notion of power indices can be used as a naive solution concept for the game itself. One can postulate that, if no specific payoff allocation rule is specified a priori, then normalized power indices can be used to allocate payoffs. Alternatively, it can be used for payoff allocation when there is a common pool in which the game players share their resources and there is no different coalition formation process by the players themselves.

Widely used power indices are the *Shapley-Shubik Power Index* (abbreviated here as SSPI) [9], and the *Banzhaf Power Index* (abbreviated here as BPI) [10]. These are generally sums of the *marginal contributions* $(v(S) - v(S \setminus \{i\}))$ of a player i to each coalition, where $v(S)$ is the characteristic function of the game, weighted by different probability distributions over the set of coalitions. We say that player i is *critical* to coalition S if its marginal contribution is 1, otherwise it is non-critical. Formal definitions of the indices appear next.

The SSPI assumes all permutations of the order that members form a coalition are equally likely and is defined by:

Definition 5.4.1 (SSPI).

$$SSPI_i(N, v) = \sum_{\substack{S \subseteq N \\ (S \ni i)}} \frac{(|S| - 1)!(N - |S|)!}{N!} (v(S) - v(S \setminus \{i\})) \tag{5.1}$$

The BPI assumes, on the other hand, that all possible coalitions that contain i are equally likely and is defined by:

Definition 5.4.2 (BPI).

$$BPI_i(N, v) = \frac{1}{2^{N-1}} \sum_{\substack{S \subseteq N \\ (S \ni i)}} (v(S) - v(S \setminus \{i\})), \tag{5.2}$$

for $i = 1, \ldots, N$.

Another popular power index based on minimal winning coalitions, i.e., it is based on the number of coalitions that satisfy the coalitional equilibrium concept of the *core* described later in the chapter, is the *Holler-Packel Power Index* (HPI) [8]. For simple games, the HPI is defined as:

Definition 5.4.3 (HPI).

$$HPI_i(N, v) = \sum_{S \in M(N,v)} (v(S) - v(S \setminus \{i\}))$$

$$= |\{S \in M(N, v) : i \in S\}|, \tag{5.3}$$

for $i = 1, \ldots, N$, *where* $M(N, v)$ *is the set of all minimal winning coalitions.*

We define also the normalized *Holler-Packel value* $\bar{HPI}_i$ = $HPI_i/\sum_{i=1}^{N} HPI_i$, which represents the proportion of minimal winning coalitions player i is in.

In the Weighted Voting Game it is more accurate to measure a player's power using either the BPI or the SSPI. The two power indices often come up with different measures of power for each player yet neither one is necessarily a more accurate depiction. Thus, which method is best for measuring power is dependent on which assumption best fits the situation. The Banzhaf measure of power is based on the idea that players are free to come and go from coalitions, negotiating their allegiance. The Shapley-Shubik measure centers on the assumption that a player makes a commitment to stay upon joining a coalition. We will later analyze in the networking scenario how these 2 power indices are not as accurate for the goal of the scenario as the Holler Packel measure is, because it is based on an equilibrium concept; furthermore we will propose a new power index by refining the HPI to better fit the specifics of our scenario.

5.5   The Coordination Game

Another game model we examine, which we will be making use of in the subsequent networking scenario is the *Coordination Game* model. In game theory, Coordination Games are a class of games with multiple pure strategy Nash equilibria in which players choose the same or corresponding strategies. In such games all players can realize mutual gains, but only by making mutually consistent decisions.

Let us consider the general payoffs of a Coordination Game, specifically a 2-player, 2-strategy game, with the payoff matrix given in Table 5.1.

Table 5.1: General payoffs of a Coordination game

	Player 2 Cooperates	Player 2 Defects
Player 1 Cooperates	A, a	C, c
Player 1 Defects	B, b	D, d

In a Coordination Game the following inequalities in payoffs hold:

- For Player 1: $A > B$, $D > C$.

- For Player 2: $a > c$, $d > b$.

We use the example here of the strategies *cooperate* and *defect* since we are presenting a 2-strategy game; however, any other strategies could be part of the Coordination Game, as long as the payoff relationship holds and as long as the same or corresponding (with same payoff relationships) strategies are played by the players in the same order. The pure Nash Equilibria in this game are the strategy profiles of mutual cooperation and mutual defection, i.e., (row 1, column 1) and (row 2, column 2). This is a 2-player, 2-strategy game but it can be extended to more than 2 strategies for each player, with Nash Equilibria found in the diagonal of the square formed from the strategies, i.e., (row 1, column 1), (row 2, column 2), (row 3, column 3), (row 4, column 4), etc., given that corresponding strategies (or same) are played by the 2 players in the same order. Similarly, a Coordination Game of more than 2 players is possible, but it is outside within the scope of this book.

A typical story for a Coordination Game is choosing the side of the road upon which to drive. It is important for drivers to be able to coordinate themselves when driving so that the common understanding of the Nash Equilibrium can be applied toward their own safety. For example, assume that 2 drivers meet in a narrow road. Both have to swerve in order to avoid a head-on collision. If both execute the same swerving maneuver, both turn toward their left hand side or both turn toward their right hand side, they will manage to pass each other. However, if one of them chooses to turn toward his left hand side and the other toward his right hand side, they will collide. In the general payoff table (Table 5.1), replace the cooperation strategy with the strategy *right* and the detection strategy with the strategy *left*. It is obvious that the safe moves and the corresponding Nash Equilibria are mutual left or mutual right, i.e., (row 1, column 1) and (row 2, column 2). Either Nash Equilibrium is a desirable outcome and any of the remaining strategy combinations is obviously not desirable at all. Next we consider some numerical values instead of the general payoffs to better understand the various payoffs in the game. Assume that successful passing of the narrow road is represented by a payoff of 10, and a collision by a payoff of 0. Therefore, the payoffs will look like Table 5.2. In this example, it doesn't matter which side both players

Table 5.2: Payoffs of a Coordination Game: Example 1

	Player 2 Right Turn	Player 2 Left Turn
Player 1 Right Turn	10, 10	0, 0
Player 1 Left Turn	0, 0	10, 10

pick, as long as they both pick the same. Furthermore, both solutions are socially efficient.

Let us now consider a not so straightforward Coordination Game, with payoffs as shown in Table 5.3. Although in this example the payoff relationships hold, it is obvious that (row 1, column 1) equilibrium dominates the equilibrium of (row 2, column 2) for both players and is thus more desirable as a game outcome. Therefore, both players prefer that they both play strategy A than both playing strategy B, in the same way that they prefer that they mutually play the same strategy than playing different strategies. The real puzzle arises in Coordination Games with payoffs as shown in Table 5.4, where there truly exist conflicting interests of the 2 players (these types of

Coordination Games are also referred to as *Conflicting Interests Coordination* games or even *Battle of the Sexes* games). In this type of Coordination Game both players prefer engaging in the same activity over going alone, but their preferences differ over which activity they should engage in. Player 1 prefers that they both play option A while Player 2 prefers that they both play option B.

Table 5.3: Payoffs of a Coordination Game: Example 2

	Player 2 A	Player 2 B
Player 1 A	10, 10	0, 0
Player 1 B	0, 0	5, 5

Table 5.4: Payoffs of a Coordination Game: Example 3

	Player 2 A	Player 2 B
Player 1 A	10, 5	0, 0
Player 1 B	0, 0	5, 10

Coordination Games also have mixed strategy Nash equilibria, which are beyond the scope of this book since we aim to find solutions that we can immediately apply to real-life situations as pure actions. Games like the driving example above have illustrated the need for solutions to coordination problems. Often we are confronted with circumstances where we must solve coordination problems without the ability to communicate with the opponent, and equilibria indicate the solution tendencies by both players. In the networking scenarios that we select in this book, we do not deal with human thought but with communication entities programmed to act in certain ways. Thus the need for an equilibrium point, on which the entities involved may be programmed to operate, is more prevalent.

5.6   Cooperation between multiple networks: Coalitions toward network synthesis

Access networks participating in a converged platform may cooperate and share resources. Cooperation between multiple access networks might be necessary in order to meet service requirements[1], e.g., when traffic forecast indicates that no single network alone can handle the anticipated demand over a certain period (days, months, etc.). Potentially, many different combinations of access networks can jointly provide sufficient resources to meet service demands. The

---

[1]Non-cooperative proposed solutions for meeting service requirements in NGN environments exist, for example see [5].

access networks participating in the selected combination are expected to receive a payment by the converged platform administrator, briefly discussed in Chapter 1.

5.6.1   Scenario overview

Consider as an example scenario of access network cooperation the need for accommodation of a multiparty service in the case that a number of collocated users (e.g., participating in a conference) have subscribed for the same live interactive multimedia multiparty service, with its starting time and duration known in advance. Each of the subscribed users have the same interfaces to the access networks participating in the converged network in their specific location (e.g., WiFi, WiMax and WCDMA).

The particular scenario considers that we have the case that none of the participating access networks can alone support all the users subscribed to the multiparty service. Then, network cooperation can ensure a more efficient network resource planning and service support to all users, by having the networks forming coalitions and thus collectively managing to provide the total resources necessary to support the service.

A user or application will be indifferent to this cooperation, as long as its QoS requirements are satisfied. On the other hand, the networks' incentive to participate in such coalitions is the service payment, which will be paid to the *winning* coalition. Therefore, the payoff allocation to the members of the *winning* coalition is also consideration for the networks participating in the coalition formation process. Given that each network participating in a coalition is solely motivated by its need to maximize its revenues from the particular service, this chapter investigates how such coalitions can be formed that satisfy the multiparty service subscribers.

5.6.2   Network Synthesis Game

Consider that the different access networks are under different administration authority or ownership and therefore, the decision of whether to participate in a certain resource combination (or network coalition) or not, would be shaped by the access network's goal to maximize its revenue from the contributed resources. As a result, a coalition game would arise in this environment, with each access network aiming at participating in a "prevailing" coalition (the one that will win) and at yielding the largest possible benefit to itself. The formation of coalitions depends on the available resources of each access network, as well as on the way payoffs are allocated to the participating networks. In this section, we introduce this coalition game and refer to it as the *Network Synthesis* game.

The players participating in the Network Synthesis Game aim to maximize their payoff by participating in the winning coalition. We consider for this game, a payoff allocation approach according to values of (normalized) power indices, i.e., numerical values that are used to measure the influence of a player on the formation of coalitions and thus on the game itself. Payoffs are thus determined based on the power of each access network in the game, i.e., its index. We consider well-known indices, such the *Shapley-Shubik* index [9], *Banzhaf* index [10] and the *Holler-Packel* index [8], described earlier. We proceed to propose a new index, called the *Popularity Power* index, which associates the popularity of each access network to the number of stable coalitions it participates in. This new index aims to achieve fairness, in the sense that it only considers the possible coalitions that would be formed if payoffs were assigned proportionally to the players' contributions, i.e., in a fair manner. Accordingly, we evaluate the proposed power index by introducing an analysis of the stability of coalitions according to the *core* and *inner core* concepts [4], considering both transferable and non-transferable payoffs, and we show that the coalitions that would be formed in the case of the *Popularity Power* index are only coalitions that are stable.

Definition of the Network Synthesis Game

Definition 5.6.1 (The Network Synthesis Game). *The Network Synthesis Game is described as follows. Let $\mathcal{N} = \{1, 2, \ldots, N\}$ denote the set of players (access networks) and S the set of all possible coalitions, i.e., the set of all non-empty subsets of $\mathcal{N}$. Let B denote the least amount of resources needed for accommodating service demands, and let b_i denote the amount of available resources of the ith member of a coalition (members can be ordered arbitrarily). It will be assumed that the available resources of each member are known to all members of a coalition[2].*

The characteristic function of the game is

$$v(S) = \begin{cases} 1, & \text{if } \sum_{i=1}^{|S|} b_i \geq B \\ 0, & \text{otherwise}. \end{cases} \tag{5.4}$$

That is, a coalition has positive value only if the sum of available resources of its members is greater or equal to the resource threshold B. This definition corresponds to a simple (or 0-1) game; the game is also monotonic since $v(S_1) \leq v(S_2)$ for all $S_1 \subseteq S_2$. A coalition S is said to be *winning* if $v(S) = 1$, otherwise it is said to be *losing*. A player $i \in S$ is said to be a *null player for coalition S* if $v(S) = v(S \setminus \{i\})$. It is generally called a *null player* if this holds for every coalition S to which it may belong. To avoid trivialities, we will generally assume that $b_i < B \; \forall i \in \mathcal{N}$, and that $\sum_{i=1}^{N} b_i \geq B$.

Candidate solutions to the game are the so-called *minimal winning coalitions*. A winning coalition is said to be minimal if it becomes a losing one

---

[2]In any wireless environment there are certain constraints such as signal interference and user mobility that make the estimation of available resources a difficult task and furthermore, the competing nature of the participating access networks may urge them to withhold or distort this information. It will be assumed that appropriate mechanisms and policies are in place that make the available resources vector $(b_1, \ldots, b_N)$ known to all operators.

upon departure of any member. A related notion is that of a *by-least winning coalition*[3]. Denoting by $W(S) = \sum_{i=1}^{|S|} b_i$ (the sum of the available resources of the members of S), a coalition S is said to be *by-least winning with* (or *for*) player i, if it is a minimal winning coalition, it contains i, and for any other minimal winning coalition S' containing i it holds that $W(S) \leq W(S')$.

We will be using the notion of the power indices to resolve our game described earlier in the chapter. We will look at the properties of the various power indices and determine which is more suitable to consider in our game. For example, an important property of both the SSPI and the BPI indices in monotonic simple games is that players with greater weight (contribution) also get a greater index. This is evident here since if a player i is critical to a coalition $S \cup \{i\}$, then a player i' with $b_{i'} > b_i$ is also critical to coalition $S \cup \{i'\}$. This is also referred to as "monotonicity of the players' power indices to the weights."

Recall that the idea of the power discussed originates from the Weighted Voting Game corresponding to the number of votes corresponding to each voter in an election system, where each voter carries a different voting power. In the following section, we show the relationship between the Weighted Voting Game and the Network Synthesis Game, justifying the study of power indices as a way to resolve the Network Synthesis Game.

Equivalence to the Weighted Voting Game

This section demonstrates the equivalence of the Network Synthesis Game to the Weighted Voting Game, a well-studied paradigm from which many useful conclusions can directly apply.

Definition 5.6.2. *A Weighted Voting Game consists of N players and a weight vector $w = (w_1, w_2, \ldots, w_N)$, where w_i reflects the "voting weight" of player i. Let $W = \sum_{i=1}^{N} w_i$. For a coalition S, the characteristic function of the game is*

$$v(S) = \begin{cases} 1, & \text{if } \sum_{i=1}^{|S|} w_i > \frac{W}{2} \\ 0, & \text{otherwise} . \end{cases} \tag{5.5}$$

We assume $w_i \leq W/2 \; \forall i \in \mathcal{N}$.

Proposition 5.6.1. *The Weighted Voting Game can be mapped into a Network Synthesis Game with networks' resources $b_i = w_i$ and minimum required resources to accommodate a service equal to $B = \frac{W}{2}$.*

Proof. To prove the equivalence, it suffices to show that there exists a one-to-one mapping between vectors w and $b = (b_1, \ldots, b_N)$ so that $\sum_{i \in S} w_i > W/2$ if and only if $\sum_{i \in S} b_i \geq B \; \forall S \subseteq \mathcal{N}$. A mapping satisfying these requirements

---

[3]In [3], the authors define a coalition S to be "least-winning" if it is minimal winning and for any other minimal winning coalition S' it holds that $W(S) \leq W(S')$. We introduce a related definition here from the point of view of each player $i \in S$.

is readily obtained by setting $w_i = b_i/2$ and W to a number B^* arbitrarily close to B such that $\sum_{i \in S} b_i \geq B$ iff $\sum_{i \in S} b_i > B^*$. (It is straightforward that such a number exists, since we have discrete b_i values.)   □

Furthermore, based on the equivalence of the Network Synthesis Game to the Weighted Voting Game, two useful observations made in [1, 3] about the behavior of the power indices defined previously, can be transferred here: first, that restricting our attention to minimal winning coalitions as with the HPI results in weaker players (in our case, players with relatively smaller available resources) getting higher power, compared to the measurement with the SSPI and the BPI. Secondly, that with the HPI the monotonicity of the players' power indices to their weights may not be preserved: a player with smaller weight may get a higher HPI ranking than a player with greater weight.

Payoff Allocation

In the Network Synthesis Game, we seek the most possible winning coalition to support a particular traffic demand, i.e., the coalition that is most likely to be formed, so that the allocation of payoffs is more fair. Access networks participate in a coalition and offer their resources in return for some revenue (payoff). For example, the converged platform administrator may have a fixed amount of money to distribute to access networks in a coalition, as a reward for reserving their resources to handle the specific service(s). It is considered that all networks are independent and behave rationally, and that the objective of each network is to maximize its payoff.

Clearly, which coalition(s) will finally be formed depends on how payoffs are allocated. Moreover, if coalitions are formed arbitrarily and each access network could participate in more than one coalition, then the criterion for selecting which coalition to participate in, is the payoff received from each, naturally preferring the highest. This preference should be captured in the power index used to allocate payoffs, i.e., the power index should be defined in a way that the most *popular* coalition, i.e., the one preferred by most players, is favored. Let's concentrate on payoffs and consider payoff allocations for the game without the use of power indices.

Assuming that no power indices are used, and that payoffs are assigned based on coalition formation, let the total payoff allocated to the set of players be P and the payoff allocation vector be $p = (p_1, p_2, \ldots, p_N)$, such that $p_i \geq 0$ $\forall i = 1, \ldots, N$ and $\sum_{i=1}^{N} p_i = P$; an allocation satisfying the above conditions is said to be *feasible*.

We consider two cases: a) *transferable* payoffs between the access networks, and b) *non-transferable* payoffs. In the transferable payoff case, individual access networks can transfer any portion of their payoff to other members of the coalition, as long as their final payoff remains greater than zero. These transfers may be viewed as side-payments, used as a means to "attract" other players in a specific coalition. In the non-transferable case, such side payments are not allowed, and we will consider that access networks attain a payoff that is proportional to their resource contribution, relative to the other members of the coalition. More specifically, if this winning coalition is $\mathcal{K}$ consisting of $K \leq N$ access networks with available resources $b_1, b_2, \ldots, b_K$, then

$$
p_i = \begin{cases} \frac{b_i}{\sum_{j=1}^{K} b_j} P, & \text{if } i \in \mathcal{K} \\ 0, & \text{otherwise.} \end{cases} \tag{5.6}
$$

(It holds that $\sum_{j=1}^{K} b_j \geq B$.)

When payoffs are transferable, this leads to trivial solutions of minimal-sized coalitions. When payoffs are non-transferable, the proportional payoff allocation case is more interesting and requires the notion of a "by-least winning" coalition. Given the defined power indices, we may relate HPI to minimal-winning coalitions, however, none of the defined power indices relates directly to the concept of by-least winning coalition; both concepts are defined in section 5.6.2. Consequently, in Section 5.6.3, we propose a new power index, called *Popularity Power* index (PPI), that satisfies this requirement, and further we provide a stability analysis to show that the PPI relates to coalitions that are stable.

5.6.3   A new power index

We would like to determine a *fair* way of allocating payoffs to a subset of the access networks, so that a stable coalition is formed and the desirable service(s) are provided. One could argue that every minimal winning coalition could potentially be a solution of the game. However, simple arguments show that the set of possible solutions can be further reduced, since not all minimal winning coalitions are equally likely. Rather, each player has specific preferences, i.e., to end up with higher payoff, and to be in one or more coalitions, which are by-least winning with it.

Let $M(N, v)$ denote the set of all minimal winning coalitions and let $Z_i(N, v)$ be the set of coalitions which are minimal in size for player i. A coalition S is said to be *minimal in size for* $i \in S$, iff $|S| \leq |S'|$ $\forall S' \ni i$, where $S, S' \in M(N, v)$. Considering the proportional allocation rule, if a player i belongs to two minimal winning coalitions S and S', then $(b_i / \sum_{j \in S} b_j) P > (b_i / \sum_{j \in S'} b_j) P$ if $\sum_{j \in S} b_j < \sum_{j \in S'} b_j$, and hence it would get a higher payoff in the by-least winning coalition. We denote by $L_i(N, v)$ the set of all by-least winning coalitions for player i.

The new power index, is based on the popularity of all coalitions which are in $\cup_{i=1}^{N} L_i(N, v)$ (a subset of $M(N, v)$). For each minimal winning coalition $S \in M(N, v)$, we define as its *preference index* $\omega(S)$ the total number of preferences it gathers by all players:

$$
\omega(S) = |\{i \in \mathcal{N} : S \in L_i(N, v)\}| . \tag{5.7}
$$

We define a new index, which we will call the *Popularity Power Index* (PPI), as

$$PPI_i(N,v) = \sum_{S \in M(N,v)} \frac{\omega(S)}{\sum_{k \in M(N,v)} \omega(k)} I_{iS} , \qquad (5.8)$$

where I_{iS} equals 1 if $i \in S$ and 0 otherwise. In plain words, the index PPI_i equals the probability that, if we were to pick a coalition by asking one player in $\mathcal{N}$ randomly (and further, if when this player had multiple equal preferences, he would select one of them with equal probability), then a winning coalition would be selected that contains player i. Hence, this index relates the popularity of minimal winning coalitions a player belongs in, to this player's power. As with the other indices, we can also define a normalized form of this index: $\overline{PPI_i} = PPI_i / \sum_{i=1}^{N} PPI_i$.

In the next few paragraphs, we relate PPI to the stable coalitions formed in case no power indices are used, and payoffs may be either transferrable or non-transferrable and proportionally allocated.

Stability Analysis

In this section, the well-known concept of the *core* (see, e.g., [4]) is used to examine the stability of coalitions formed according to the payoff allocations considered if no power index is used to allocate payoffs. Furthemore, we will show the equivalence of the proposed power index to payoff allocation for stable coalitions.

Descriptively, a payoff allocation to a set of N players is in the core of a coalitional game if there is no other coalition wherein each member can get a strictly higher payoff than dictated by the allocation. Such an allocation, as well as the coalitions that it induces, can be called stable, since there would not be a consensus to break these coalitions and form other ones.

In order to apply the core concept, we slightly modify the characteristic function of the Network Synthesis Game, found in Equation (5.4) in the transferable payoff case to the following:

$$v(S) = \begin{cases} P, & \text{if } \sum_{i=1}^{|S|} b_i \geq B \\ 0, & otherwise. \end{cases} \qquad (5.9)$$

That is, when the minimum necessary resources are available, the value of the characteristic function equals the total payoff.

For an allocation to be in the core of the transferable-payoff game — since we are not interested in how the payoff is divided among the members of the coalition — it is required that it does not have an incentive to deviate and obtain an outcome better for all its members.

Definition 5.6.3. *An allocation $p = (p_1, p_2, \ldots, p_N)$ is said to be in the core of the access Network Synthesis Game with transferable payoffs iff*

$$\sum_{i \in \mathcal{N}} p_i = P \text{ and } \sum_{i \in S} p_i \geq v(S), \forall S \subseteq \mathcal{N} .$$

Since $v(S)$ takes either the value 0 or P in our game, this trivially reduces to the requirement that for every winning coalition that could be formed by the networks, the sum of payoff allocations should always equal P.

For the non-transferable payoff case, we have the following definition:

Definition 5.6.4. *An allocation* $p = (p_1, p_2, \ldots, p_N)$ *is said to be in the core of the access Network Synthesis Game with non-transferable payoffs iff* $\sum_{i \in \mathcal{N}} p_i = P$ *and there exists no other payoff allocation* $y = (y_1, y_2, \ldots, y_N)$ *derived according to equation (5.6) for which* $y_i > p_i$, $\forall i \in S \subseteq \mathcal{N}$, *for any* $S \subseteq \mathcal{N}$.

That is, the allocation must be feasible and there should exist no other allocation which gives strictly higher payoff to all members of a coalition.

A single winning coalition can be mapped to an allocation in the core. In the non-transferable payoff case, this is defined to be the coalition $\mathcal{K}$, based on which the payoff vector is derived. In the transferable payoff case, this is defined as $\{i \in \mathcal{N} : p_i > 0\}$, the set of players with positive payoff. We can then speak about "coalitions in the core," as the set of winning coalitions for which their corresponding allocations are in the core.

Not all winning coalitions are in the core. In fact, we have the following:

Theorem 5.6.1. *In both the transferable and non-transferable payoff cases defined above, only minimal winning coalitions are in the core.*

Proof. In the transferable payoff case, notice that for any non-minimal winning coalition, a corresponding minimal one can be formed by the players that are non-null (in the coalition). Then for any payoff allocation to players in the non-minimal winning coalition, the players in the minimal one can divide the excess payoff in such a way that they all get strictly higher payoff. In the non-transferable payoff case, the statement of the theorem follows directly from equation (5.6). □

Remark 1. *This theorem is an adaptation of Riker's size principle [2] to this game, which was also shown for Weighted Voting Games in [1].*

Hence it is reasonable to direct our attention to minimal winning coalitions, $M(N, v)$.[4] In the case of transferable payoffs, only minimal winning coalitions which are also minimal in size for at least one of their members, denoted as $Z_i(N, v)$, will be in the core. In a sense, when we have transferable payoffs, all minimal winning coalitions of the same size must be treated as equivalent. In non-transferable payoff games, only coalitions that are by-least winning for at least one player, denoted by $L_i(N, v)$, or simply by L_i, are solutions in the core of the game.

---

[4]Although the characteristic function v is defined usually for transferable payoff games, it is also used in this thesis in notations concerning the non-transferable payoff game such as $M(N, v)$ and $L_i(N, v)$, since, along with (5.6), it can be used to define the latter game.

We further proceed to study which coalitions are in the *inner core* of the game. The inner core [4] is a subset of the core that contains coalitions that are "more stable," in the sense that there exists no randomized plan that could prevent their formation.

Definition 5.6.5. *A randomized plan is any pair* $(\eta(S), y(S))$, $S \subseteq \mathcal{N}$, *where* η *is a probability distribution on the set of coalitions, and* y *is the vector of payoff allocations for the members of coalition* S, $y(S) = (y_i(S))_{i \in S}$. *For nontransferable payoff games, the inner core is composed of all allocations* p (*or corresponding coalitions*) *for which* $\sum_{S \supseteq \{i\}} \eta(S) y_i(S) < \sum_{S \supseteq \{i\}} \eta(S) p_i$, *for some* $i \in \mathcal{N}$, *in all randomized plans* $(\eta(S), y(S))$.

The inner core concept is used in cases where there is a mediator that invites individual players to form a coalition which is not known deterministically, but only with a certain probability distribution. Consider for example that in the Network Synthesis Game, the converged platform administrator (mediator) informs the networks that, in case they don't arrive to an agreement by themselves, then a coalition of its choice will be selected, which will be S_1 with probability p_1, or S_2 with probability $p_2 = 1 - p_1$. Then, in order for a coalition to be stable, the payoff given to each network should not be smaller than the mean payoff anticipated in the coalition of the platform administrator's choice.

Theorem 5.6.2. *In the access Network Synthesis Game with non-transferable payoffs proportional to the available resources of the participating networks, all coalitions which are by-least winning for at least one of their members are in the inner core of the game.*

Proof. In the Network Synthesis Game, if an access network participates in several coalitions that are by-least winning with it, then in the non-transferable payoff case it would get the highest possible reward in every one of these coalitions. This reward would further be the same in every randomized plan among these coalitions and lower for randomized plans containing coalitions other than the by-least winning.

Since, for a player i, the allocation in a by-least winning coalition is the maximum it can get, there exists no randomized plan that could block these coalitions from forming and hence the latter are in the inner core of the game. $\qquad \Box$

Remark 2. *Of all by-least winning coalitions which are in the inner core, we may informally state that those that are by-least winning for all their members are "most stable." This will be made formal by defining a new concept of "stability under uncertainty of formation" in Definition 5.6.6.*

A nice property of the game is formulated in the following:

Theorem 5.6.3. *In the access Network Synthesis Game with non-transferable payoffs proportional to the available resources of the participating networks,* *there exists at least one coalition which is by-least winning for all its members. Further, regardless of the payoff allocation, there exists at least one coalition that is minimal in size for all its members.*

Proof. We demonstrate the theorem only for by-least winning coalitions. The proof for minimal in size coalitions is similar since the existence of a by-least winning coalition by definition implies the existence of at least a minimal winning coalition.

The proof is straightforward when there exists a coalition S, such that $\sum_{i \in S} b_i = B$, since then S is by-least winning for all its members. When $\sum_{i \in S} b_i > B$ for some $S \in \cup_{i=1}^{N} L_i$, and there exists $j \in S$ such that $S \notin L_j$, then necessarily another coalition $S_1 \neq S$ exists such that $S_1 \in L_j$ and $\sum_{i \in S_1} b_i < \sum_{i \in S} b_i$. Similarly now, if there exists $k \in S_1$, $k \neq j$, such that $S_1 \notin L_k$, then there exists another coalition $S_2 \notin \{S_1, S\}$ such that $S_2 \in L_k$ and $\sum_{i \in S_2} b_i < \sum_{i \in S_1} b_i$. Continuing this procedure, since we have a finite number of players, a finite sequence of coalitions $S, S_1, S_2, \ldots, S_m$ is produced, for which $\sum_{i \in S} b_i > \sum_{i \in S_1} b_i > \sum_{i \in S_2} b_i > \cdots > \sum_{i \in S_m} b_i > B$ and S_m is by-least winning for all its members. □

Hence, in both transferable and non-transferable payoff cases studied, interests of at least some players coincide. However, it will be reasonable to assume that a player will prefer a minimal winning coalition even though it would not be by-least winning with it, if otherwise it would be excluded from the prevailing coalition and receive zero payoff.

The Coordination Game

We have established that each access network i would maximize its payoff and hence prefer one of the coalitions in $Z_i(N, v)$ (transferable payoff case) or $L_i(N, v)$ (proportional payoff case) to eventually be formed. Unless $\cap_{i=1}^{N} Z_i \neq \emptyset$ in the former, or $\cap_{i=1}^{N} L_i \neq \emptyset$ in the latter case, there is no mutually preferred coalition, and we are led to a *Coordination Game* where at least one player has *conflicting preferences* with one or more of the others.

A possible resolution of such a Coordination Game follows, which is based on the idea of calculating the probability that these coalitions would randomly form. The analysis applies equally to the transferable and non-transferable payoff cases, and in what follows we shall use $\mathcal{G}_i(N, v)$ (or simply $\mathcal{G}_i$) to denote either $Z_i(N, v)$ or $L_i(N, v)$, depending on which case we study.

Resolution of the Coordination Game

The ultimate goal of the game is to find out if one or more stable coalitions can form, so that the service demands are satisfied. Further, what is more important is that, under the proportional payoff allocation rule, if an access network i can be in multiple minimal winning coalitions, then it prefers the one (or ones) which is (or are) by-least winning with it. (Clearly, if there are

more than one by-least winning coalitions for a player, then they must sum
to the same total resource amount.) For each player i, we denote by $L_i(N, v)$
the set of all by-least winning coalitions for i (this set will also be denoted
simply by L_i).

For cases where some players have more than one by-least winning coalitions, we may examine which coalitions are more likely to be formed. For
$i = 1, \ldots, N$, we define the *probability of formation* $P_f^{(i)}(S)$ to be the probability that player i would anticipate coalition S to be formed, if player i
participated in it and all other players $j \in S$, $j \neq i$ would independently
choose to participate in one of their preferred coalitions with equal probability. That is,

$$P_f^{(i)}(S) = \begin{cases} \prod_{\substack{j \in S \\ j \neq i}} \frac{1}{|\mathcal{G}_j|}, & \text{if } S \in \bigcap_{j \in S, j \neq i} \mathcal{G}_j \\ 0, & \text{otherwise.} \end{cases} \tag{5.10}$$

Then it is reasonable that a player would ultimately prefer to participate in
the coalition which has the highest probability of formation, and hence, under
such uncertainty, would offer it the greatest expected payoff.

To appoint a name to this concept of stability, we define a coalition S
to be *stable under uncertainty of formation* if and only if there exists no
other coalition S' with a common member with S that anticipates a higher
probability of formation for S'.

Definition 5.6.6. *In the access Network Synthesis Game with transferable or
non-transferable payoffs, a coalition S is stable under uncertainty of formation
iff $P_f^{(i)}(S) > 0 \; \forall i \in S$ and*[5]

$$\nexists S' \neq S \; s.t \; P_f^{(j)}(S') > P_f^{(j)}(S) \; for \, j \in S \cap S'.$$

This notion would help to refine solutions of the Coordination Game. It
also creates a formal ground to confine solutions to coalitions which are minimal in size (transferable payoff case) or by-least winning (proportional payoff
case) for all their members: by definition, only such coalitions are stable under uncertainty of formation (in view of (5.10), other coalitions have zero
probability of formation for at least one member).

Remark 3. *PPI associates a player's value to the number of stable coalitions it participates in; the higher this number, the greater the index value.
In fact the preference denoted by $\omega(S)$, in the definition of PPI, refers to
how many players consider coalition S, by-least winning. Clearly, the coalition with the highest ωS is stable under uncertainty of formation, according
to Definition 5.6.6.*

Thus, PPI excludes a number of coalitions which are not stable and hence
would not appear if access networks formed coalitions independently. In addition, the PPI is more fair than other indices examined, in the sense that it only

---

[5]The term s.t means "such that."

considers stable coalitions in the inner core that would be formed if payoffs were allocated proportionally to players' contributions, i.e., in a fair manner. These coalitions have been shown to be the most probable to be formed, both in the transferrable and in the non-transferrable payoff cases. Section 5.6.4 undertakes the evaluation of the various indices and demonstrates how the PPI allocates payoffs only to the players that would participate in the minimal in size or by-least winning coalition, i.e., players that are in the inner core of the game, as we have shown that all coalitions which are by-least winning for at least one player are in the inner core.

5.6.4 Evaluating the game

This section examines the numerical behavior of all power indices described in the previous section.

Power index values are examined for different numbers of players and different distributions of available resources. Even though the theory extends to many players, we have selected game instances of a few players only (3, 4 and 5 players) for illustrative purposes. Individual resources for each test instance presented sum up to the same total resource amount; this is done in order to better compare results between the different cases for the same number of players. Note that collaboration between players are considered only if the sum of their resources adds up to or exceeds the amount of available resources, i.e., one (1) in the simulations.

For each set of players we consider different distributions D_i ($i = 0, \ldots, 5$) of available resources, from the case $i = 0$ where resources are uniformly distributed between access networks, i.e., each network has the same fraction of available resources, to non-uniform cases ($i = 1, \ldots, 5$), carefully selected to exhibit the varying allocations of the power indices, carefully selected to exhibit the values of the power indices when available resources are about the same, or resources are concentrated in only a few of the networks.

To avoid taking absolute values, we have considered available resources of each player i normalized with respect to the minimum resource requirement B, i.e., b_i/B. For each of the power indices BPI, SSPI, HPI and PPI, we examine both the values of the indices as well as their rankings.

The values of normalized available resources are shown in Tables 5.5, 5.7, and 5.9, for the cases of 3, 4, and 5 players respectively. For each one of these cases the 3 power indices are generated also in a normalized form so that they add up to one. The values of indices for these cases are shown in Tables 5.6, 5.8, and 5.10.

Differences between the indices' values become more pronounced as the number of players increases (the increased number of possible coalitions allows such differences to show). In general, the SSPI and BPI give similar values, favoring the players with greater available resources. (A closer inspection reveals that SSPI systematically does that to a slightly greater extent than BPI.) On the other hand, the HPI and PPI give a higher power to

Table 5.5: Instance 1: 3 players

Distribution	$\frac{b_1}{B}$	$\frac{b_2}{B}$	$\frac{b_3}{B}$
D_0	0.4	0.4	0.4
D_1	0.8	0.2	0.2
D_2	0.8	0.3	0.1
D_3	0.9	0.2	0.1
D_4	0.6	0.35	0.25
D_5	0.55	0.45	0.25

Table 5.6: Instance 1 indices

Distribution	Index	Player 1	Player 2	Player 3
D_0	BPI	0.33	0.33	0.33
	SSPI	0.33	0.33	0.33
	HPI	0.33	0.33	0.33
	PPI	0.33	0.33	0.33
D_1	BPI	0.6	0.2	0.2
	SSPI	0.67	0.17	0.17
	HPI	0.5	0.25	0.25
	PPI	0.5	0.25	0.25
D_2	BPI	0.5	0.5	0
	SSPI	0.67	0.33	0
	HPI	0.5	0.5	0
	PPI	0.5	0.5	0
D_3	BPI	0.6	0.2	0.2
	SSPI	0.67	0.17	0.17
	HPI	0.5	0.25	0.25
	PPI	0.5	0	0.5
D_4	BPI	0.33	0.33	0.33
	SSPI	0.33	0.33	0.33
	HPI	0.33	0.33	0.33
	PPI	0.33	0.33	0.33
D_5	BPI	0.5	0.5	0
	SSPI	0.67	0.33	0
	HPI	0.5	0.5	0
	PPI	0.5	0.5	0

Table 5.7: Instance 2: 4 players

Distribution	$\frac{b_1}{B}$	$\frac{b_2}{B}$	$\frac{b_3}{B}$	$\frac{b_4}{B}$
D_0	0.4	0.4	0.4	0.4
D_1	0.85	0.25	0.25	0.25
D_2	0.8	0.55	0.15	0.1
D_3	0.95	0.45	0.1	0.1
D_4	0.6	0.4	0.35	0.25
D_5	0.55	0.5	0.3	0.25

Table 5.8: Instance 2 indices

Distribution	Index	Player 1	Player 2	Player 3	Player 4
D_0	BPI	0.25	0.25	0.25	0.25
	SSPI	0.25	0.25	0.25	0.25
	HPI	0.25	0.25	0.25	0.25
	PPI	0.25	0.25	0.25	0.25
D_1	BPI	0.7	0.1	0.1	0.1
	SSPI	0.75	0.083	0.083	0.083
	HPI	0.5	0.17	0.17	0.17
	PPI	0.5	0.17	0.17	0.17
D_2	BPI	0.5	0.3	0.1	0.1
	SSPI	0.75	0.17	0.042	0.042
	HPI	0.4	0.2	0.2	0.2
	PPI	0.33	0	0.33	0.33
D_3	BPI	0.7	0.1	0.1	0.1
	SSPI	0.75	0.083	0.083	0.083
	HPI	0.5	0.17	0.17	0.17
	PPI	0.5	0	0.25	0.25
D_4	BPI	0.33	0.33	0.17	0.17
	SSPI	0.33	0.33	0.17	0.17
	HPI	0.25	0.25	0.25	0.25
	PPI	0.2	0.4	0.2	0.2
D_5	BPI	0.33	0.33	0.17	0.17
	SSPI	0.33	0.33	0.17	0.17
	HPI	0.25	0.25	0.25	0.25
	PPI	0.2	0.4	0.2	0.2

Table 5.9: Instance 3: 5 players

Distribution	$\frac{b_1}{B}$	$\frac{b_2}{B}$	$\frac{b_3}{B}$	$\frac{b_4}{B}$	$\frac{b_5}{B}$
D_0	0.3	0.3	0.3	0.3	0.3
D_1	0.7	0.2	0.2	0.2	0.2
D_2	0.85	0.2	0.15	0.15	0.15
D_3	0.9	0.25	0.15	0.15	0.05
D_4	0.5	0.3	0.3	0.2	0.2
D_5	0.45	0.35	0.3	0.25	0.15

Table 5.10: Instance 3 indices

Distribution	Index	Player 1	Player 2	Player 3	Player 4	Player 5
D_0	BPI	0.2	0.2	0.2	0.2	0.2
	SSPI	0.2	0.2	0.2	0.2	0.2
	HPI	0.2	0.2	0.2	0.2	0.2
	PPI	0.2	0.2	0.2	0.2	0.2
D_1	BPI	0.48	0.13	0.13	0.13	0.13
	SSPI	0.8	0.05	0.05	0.05	0.05
	HPI	0.33	0.17	0.17	0.17	0.17
	PPI	0.33	0.17	0.17	0.17	0.17
D_2	BPI	0.79	0.05	0.05	0.05	0.05
	SSPI	0.8	0.05	0.05	0.05	0.05
	HPI	0.5	0.125	0.125	0.125	0.125
	PPI	0.5	0	0.17	0.17	0.17
D_3	BPI	0.54	0.15	0.15	0.15	0
	SSPI	0.8	0.067	0.067	0.067	0
	HPI	0.5	0.17	0.17	0.17	0
	PPI	0.5	0	0.25	0.25	0
D_4	BPI	0.31	0.22	0.22	0.125	0.125
	SSPI	0.33	0.18	0.18	0.18	0.15
	HPI	0.33	0.2	0.2	0.13	0.13
	PPI	0.33	0.17	0.17	0.17	0.17
D_5	BPI	0.3	0.22	0.22	0.22	0.04
	SSPI	0.33	0.18	0.18	0.18	0.15
	HPI	0.23	0.23	0.23	0.23	0.08
	PPI	0.33	0	0.33	0.33	0

relatively weaker players. This is because weaker players have smaller contributions and hence are more often found in coalitions which are minimal winning, or by-least winning for some players.

The HPI and PPI are more appropriate indices for the game, since they exclude a number of coalitions which are not stable and hence would not appear if access networks formed coalitions independently. Comparing these two indices, we can argue that the PPI is more fair, in the sense that it only considers stable coalitions in the inner core that would be formed if payoffs were allocated proportionally to players' contributions, i.e., in a fair manner. In Tables 5.6, 5.8, 5.10, we may note that often a player is allocated zero payoff with the PPI, whereas non-zero with the HPI. These are cases where this player participates in minimal winning coalitions, but not in any of the by-least winning coalitions. For example, in Table 5.6, case D_3, Player 2 is allocated 0.25 value with the HPI, whereas 0 with the PPI, since it does not participate in the by-least winning coalition. The monotonicity of index values to players' contributions also does not hold for the PPI in this example.

5.7 Conclusions on coalitional games

This chapter investigates cooperation between multiple selfish entities through the formation of coalitions where there exists a collective goal, unlike previous types of games where the players' selfish goals have been studied instead. This is given by the utility functions for the players in previous games, which represented their individual *satisfaction* goal. These functions have been replaced by the game characteristic function in this type of coalitional games, where this characteristic function represents a collective goal. The collective goal is to participate in a coalition that can achieve the highest payoff and thus benefit indirectly from the participation, keeping in mind that it is important to participate in a successful coalition. The main strategical question is how to decide which group to join, or in the case groups are formed from scratch, how to form these groups. The chapter demonstrated the use of power indices in this process of coalition formation as well as the use of different payoffs in different situations. For the scenario demonstrated, a new power index is defined and used, urging the reader to further expand on this idea of modifying existing power indices and even defining new ones to fit the needs of specific scenario models.

References

[1] S. J. Brams and P. C. Fishburn, *When is size a liability? Bargaining power in minimal winning coalitions*, C.V. Starr Center for Applied Economics, New York University: Working Papers, no. 94-07, 1994, http://ideas.repec.org/p/cvs/starer/94-07.html

[2] W. H. Riker, *The Theory of Political Coalitions*, Yale University Press, 1962.

[3] S. J. Brams and P. C. Fishburn, *Minimal winning coalitions in weighted-majority voting games*, Social Choice and Welfare, v. 13, no. 4, 1996, pp. 397-417, http://ideas.repec.org/a/spr/sochwe/v13y1996i4p397-417.html

[4] R. B. Myerson, *Game Theory: Analysis of Conflict*, Harvard University Press, September, 1991.

[5] K. M. Rubaiyat and A. Jamalipour, *Lossy utility based outage compensation in Next Generation Networks*, Innovations in NGN: Future Network and Services, 2008. K-INGN 2008. First ITU-T Kaleidoscope Academic Conference, May 2008, pp. 389–396, doi:10.1109/KINGN.2008.4542292

[6] J. C. Harsanyi, *Measurement of Social Power, Opportunity Costs, and the Theory of Two-Person Bargaining Games*, Behavioral Science, v. 7, 1962, pp. 67–81.

[7] J. C. Harsanyi, *Games with Incomplete Information Played by Bayesian Players, Parts I, II, and III*, Behavioral Science, v. 14, 1967, pp. 159–182, 320–334, 486–502.

[8] R. Haradau and S. Napel, *Holler-Packel value and index: A new characterization*, Homo Oeconomicus, v. 24, no. 2, 207, pp. 255-268.

[9] L. S. Shapley, and M. Shubik, *A Method for Evaluating the Distribution of Power in a Committee System*, The American Political Science Review, v. 48, no. 3, pp. 787-792, http://www.jstor.org/stable/1951053, American Political Science Association, 1954.

[10] J. F. Banzhaf, *Weighted voting doesn't work: A mathematical analysis*, Rutgers Law Review, v. 19, no. 2, pp. 317–343, 1965.

Chapter 6

MATLAB implementation: Strategies for Iterated Prisoner's Dilemma type of games

6.1 Introduction

In this chapter we provide guidelines as to how one can implement an iterated game simulation, in order to evaluate the selected player strategies. Toward this end, we demonstrate the implementation of the Iterated Prisoner's Dilemma, discussed in previous chapters, and explain how the reader can extend this demonstrative implementation to include his/her own strategies for this particular type of game. The code for these examples, as well as example runs can be found in *http://www.NetRL.cs.ucy.ac.cy* [1].

The implementation of the Iterated Prisoner's Dilemma strategies is done in MATLAB [2] and is based on a publicly available implementation of the Iterated Prisoner's Dilemma found in [3]. For the purposes of the examples in this chapter, we have used MATLAB version R2011a and this is installed in a Windows 7 operating system. Since different versions of MATLAB, and different implementations are available, some minor code adjustments might be necessary.

In this chapter we provide guidelines on how to use this framework for the purpose of implementing the Iterated Prisoner's Dilemma strategies, or variants thereof, both in terms of structuring the code as well as in terms of programming details of the MATLAB environment. Our goal is to offer a concise and easy to follow tutorial so that the interested reader can easily reproduce and modify the code to test different kind of strategies and payoff sets.

6.2 Initializing the execution

The code that we present is organized in 4 different files, which makes it easier to manage and manipulate. The way we handle the 4 files is to assign one as the main file, the file that we primarily execute, and the *main* file then assigns control, i.e., calls for execution of a secondary file when this is necessary. The secondary files may assign control to other secondary files but not to the *main* file, which may only be called and executed by the user. We discuss the calls from one file to another, as the chapter unfolds.

At first we look at the code starting with the *main* file, which is the initial point of executing this code. We named this first file *main.m*, where the extension **.m* is given for MATLAB executable files. This file is executed within the MATLAB environment by simply typing the name of the file (without the extension) in the *Command Window* of the MATLAB workspace. If the file is not in the current workspace directory, then it can be executed by using the command $run(< path \backslash filename >)$. Therefore, the command will look like:

$>> main$

Or,

$>> run('C : \backslash Users \backslash Josephina \backslash Documents \backslash matlab - IPD \backslash main.m')$

Where the path to the file, including the folder that contains the file, is:

$C : \backslash Users \backslash Josephina \backslash Documents \backslash matlab - IPD$

and the filename is:

main.m.

In Table 6.1 the code within *main.m* is presented. The first 3 lines of code indicate the required input. We hard-code this in the file, so when we need to change the numbers we go back to the file and manually replace the existing values with the new values; this can also be made interactive by using a display statement asking the user to enter these numbers and then collecting the user's input in the appropriate variable structures (structures that are used to store different values of a specific type of data).

6.3 Fixed iterations number

The first input that is required is the number of rounds the game will consist of, i.e., the number of iterations. Table 6.1 shows that we have selected a fixed number of rounds, i.e., $R = 200$. This means that the one-shot Iterated Prisoner's Dilemma game will be repeated exactly 200 times. However, having a fixed number of iterations is not always the most desirable case. A random number of iterations is much preferred in some case, and in fact we have used such random numbers in several of the results we have presented in this book. If this is the case, and the reader wishes to randomize the number of game

Table 6.1: Code in main.m

```
1.    R=200;
2.    Q=[3,4,7,2];
3.    PD=[1,4;0,3];
4.    n=length(Q);
5.    Z=zeros(n,n);
6.    for i = 1 : n
7.    for j = 1 : n
8.    [A,B,a,b]=iteratedpd(R,Q(i),Q(j),PD);
9.    Z(i,j)=a;
10.   end
11.   end
12.   Z1
13.   for i = 1 : n
14.   Scores(i)=sum(Z(i, 1 : n));
15.   end
16.   Scores
```

iterations, this can be done, by replacing the following line of the code (Line 1):

R=200;

With the following line of code:

R=rounds(0.99654);

This randomization is used without modifications from the original code in [3], as it uses the randomization mostly used in the Iterated Prisoner's Dilemma tournaments organized by Axelrod [5]. The variable R, as previously, represents an integer number of the output of the function named *rounds*, which takes as input, indicated by the brackets, a decimal point number close to but less than one, and provides a random number of runs, as described below.

At this point we remark that MATLAB is interpretive, in the sense that it interprets and executes each line of code; once *main.m* is called by typing the appropriate command in the MATLAB *Command Window*, the lines of code start getting executed in a sequential manner, unless there is a re-assignment of flow control. Flow control changes in the case we call a function because once a function is called, MATLAB is required to firstly execute the lines of code in the function and once this is completed, continue with the sequential execution of command lines. Therefore in this particular case, calling the function *rounds* will cause the suspension of the execution in the current line of code, assign control to the function being called, execute the lines in

Table 6.2: Code in rounds.m

```
function r=rounds(p)
1.   r=1;
2.   x=rand(1);
3.   while(x < p)
4.     r = r + 1;
5.     x=rand(1);
6.   end
```

the function and then assign the result of the function in the variable R, thus completing the execution of that line of code where the execution was suspended, and allowing the flow of execution to continue sequentially to the next line of code.

6.4   Randomized iteration number

Table 6.2 illustrates the code of this randomization function, which is found in a separate file named *rounds.m*, placed in the same directory as the file *main.m*. You can also place the function in the same file, however, using different files for different functions is encouraged as it makes the overall code more easily manageable.

The definition of the function *r=rounds(p)* defines the inputs necessary when the function is called, as well as the outputs that the function will return to the command that called it. The input is given by the variable in the parentheses p, which is a decimal number less than 1, given in the command in *main.m* that called this function, whereas the output is given by the variable r, which once execution of the function is completed will contain an integer representing the number of rounds of play for the game, generated by the lines of code in the function.

Line 2 assigns an initial value equal to 1 into the variable r. Line 3 assigns a random value into variable x. This value is generated by the function *rand*, which is provided by the MATLAB environment. Therefore, control of execution is suspended so that the *rand* function can execute and return its output, which will be stored in variable x. The function *rand* produces an N-by-N matrix containing pseudorandom values, uniformly distributed, where N is the input to the function. In the code presented in Table 6.2, the call of the *rand* function returns one value in a 1-by-1 matrix, between 0 and 1, i.e., from the standard uniform distribution.

Lines $4-7$ encode an iteration loop in which the value generated by *rand* is compared to the input p, in this case the number 0.99654 is used since it is also used by Axelrod [5] in the Iterated Prisoner's Dilemma tournaments. The loop counts the number of repetitions it requires for the generated number to be greater than 0.99654, and stores this count in variable r, which in turn is returned by the function to be set as the current number of iterations.

6.5   Strategies and payoffs

Once the function *rounds* is executed, and the variable R contains the number of iterations that the game will execute, we move on to Line 2 of the code in *main.m* (Table 6.1). Line 2 defines the second input to the Iterated Prisoner's Dilemma, which represents the list of strategies as vector Q. The strategies are played by both players in a round robin manner, as we can see from the two loops in lines $6 - 11$. In the first iteration of the outer loop, Player 1 plays the first strategy in vector Q, while the inner loop repeats until Player 2 plays in turn all the strategies in vector Q. In the second iteration, Player 1 plays the second strategy against all the strategies of Player 2 and so on. The play of the game is achieved with function *iteratedpd*, which we will discuss later on in the chapter. What is important to say now is that the function returns a score a for Player 1, and a score b for Player 2 for each pair of strategies played, and that score is stored in matrix Z, which at the end of the game holds the scores for each pair of strategies. For this example we have only one strategy vector Q and only one payoff vector Z, because we demonstrate the case that both players are identical. This is not always the case, especially in scenarios with heterogeneous players with different strategies. In those situations we can define a separate strategy matrix for each player, and can consequently collect the payoff for each player in a separate matrix.

The payoff values are an additional input to the game and they are given in the 2-by-2 matrix *PD*. The four values represent the four possible payoffs in the Iterated Prisoner's Dilemma interaction. Specifically, the payoffs represent the following game instances: mutual cooperation, mutual defection, cooperating while opponent defects and defecting while opponent cooperates. The matrix *Scores* provides a list of total scores for each strategy against all strategies it played against.

6.6   Collecting cumulative payoffs

We next move on to discuss the function *iteratedpd*, which is responsible for the repetition of the one-shot Iterated Prisoner's Dilemma. Table 6.3 illustrates the code for this function, once more placed in a separate file named iteratedpd.m.

Since this function is responsible for the repetitive nature of the game, we need the number of rounds, R, discussed previously as an input. This is given in Table 6.3 by the variable r. Additional inputs to this function include *p1strat* and *p2strat*, which are the integers representing the strategies currently employed by Player 1 and Player 2, respectively, and *payoff* which is the payoff matrix defined in *main.m*.

Table 6.3: Code in iteratedpd.m

```
      function [X,Y,Xscore,Yscore]=iteratedpd(r,p1strat,p2strat,payoff)
1.    mdefect=payoff(1,1);
2.    mcoop=payoff(2,2);
3.    suckerwin=payoff(1,2);
4.    suckerlose=payoff(2,1);
5.    X=[];
6.    Y=[];
7.    Xscore=0;
8.    Yscore=0;
9.    for i = 1 : r
10.   newX=play(X,Y,p1strat);
11.   newY=play(Y,X,p2strat);
12.   X=[X,newX];
13.   Y=[Y,newY];
14.   if(newX==0)
15.   if(newY==0)
16.   Xscore=Xscore+mdefect;
17.   Yscore=Yscore+mdefect;
18.   else
19.   Xscore=Xscore+suckerwin;
20.   Yscore=Yscore+suckerlose;
21.   end
22.   else
23.   if(newY==0)
24.   Xscore=Xscore+suckerlose;
25.   Yscore=Yscore+suckerwin;
26.   else
27.   Xscore=Xscore+mcoop;
28.   Yscore=Yscore+mcoop;
29.   end
30.   end
31.   end
```

The *iteratedpd* function has a number of outputs, which include primarily two vectors, X and Y, which represent the individual actions of the players, i.e., cooperate or defect, and are used by each player to aid decision about the next action to take according to the strategy each player employs. The outputs also include *Xscore* and *Yscore*, which represent the cumulative payoffs of each Player 1 and Player 2, respectively, from the Iterated Prisoner's Dilemma for each combination of strategies.

The first few lines of this function, namely Lines $2 - 5$, assign the element in a particular position of the payoff matrix, and thus the numerical value of the payoff, to a variable whose name corresponds to the respective semantic of the particular payoff value, e.g., mutual cooperation is represented by the variable *mcoop* and the corresponding value is the element found in the second row and second column of the payoff matrix. The other variables have the following semantics: *mdefect* represents the payoff for mutual defection, *suckerwin* represents the payoff of defecting while the opponent cooperates, and *suckerlose* represents the payoff of cooperating while the opponent defects.

The next few lines serve the initialization of the outputs and are followed by the loop, which repeats the single Prisoner's Dilemma games the amount of times specified in the input parameter r. The loop firstly calls the function *play* that handles the one shot Prisoner's Dilemma game, which we will discuss in detail in the next few paragraphs. Once the one-shot Prisoner's Dilemma returns the play of the game in terms of actions that the players have selected for the current round, i.e., cooperate or defect, an *if-else*, structure follows, in which the appropriate payoff for each single iteration is selected and added to each player's payoff. The payoff for each player depends on the opponent's action, i.e., whether the opponent has chosen to cooperate or defect. The X and Y vectors hold the actions of Player 1 and Player 2, respectively. Thus, the first check in the *if-else* structure is whether Player 1 has defected, where defection is represented by a zero, or else Player 1 has cooperated, where cooperation is represented by a one but the *else* is enough to check since the X and Y vectors contain only zeros and ones. The nested *if-else* structure, checks the corresponding Player 2's action, once Player 1's action is determined and once both actions are determined the appropriate payoff for the round is added to the cumulative payoff of each player, stored in variables *Xscore* and *Yscore* for Player 1 and Player 2, respectively.

6.7 A single round of the game

The remaining function that we will discuss, which completes the pieces necessary to implement an Iterated Prisoner's Dilemma in MATLAB such that several strategies of the interacting players can be juxtaposed and evaluated, is the *play* function that is placed in a separate file named *play.m* and it is the function responsible to play the single round Prisoner's Dilemma games that make up the Iterated Prisoner's Dilemma game that we are evaluating.

Table 6.4 illustrates the code for this function.

The input to the function includes the vectors containing each player's moves in the iterated game up to the point the *play* function is called, as well as the strategy that each player has adopted for the iterated game. The play function deals only with one player at a time, therefore it is called firstly with Player 1's strategy as input and returns Player 1's current move as output, and it is subsequently called with Player 2's strategy as input and returns Player 2's current move as payoff. The function returns the value 1 if the player cooperates in the current round, or the value 0 if the player defects in the current round.

Each strategy is indicated by a number, and an *if-elseif-else* structure is used to select, the code to be executed that corresponds to each particular strategy. In the illustrated code, strategy 1 is the strategy when the player always defects, strategy 2 is when the player always cooperates, strategy 3 is the Tit-for-Tat strategy, strategy 4 is the Grim Trigger strategy, and any other number used is treated similarly in this code, as the random strategy, where the player randomly selects whether to cooperate or to defect. The function used for this randomization is the MATLAB *randn* function, without any parameters, thus returning a scalar.

Implementing an iterated strategy is relatively straightforward in most cases. For example, in the case of strategy 1 and strategy 2, since the player either always defects or always cooperates, the decision on the action to take in each round is predetermined and it does not matter how the opponent plays. Thus, in the case a player always defects, i.e., strategy 1, a value of 0 will be returned as the action to take, whereas in the case a player always cooperates, i.e., strategy 2, a value of 1 will be returned as the action to take.

Now, in the case we have strategies such as Tit-for-Tat, i.e., strategy 3, or Grim Trigger, i.e., strategy 4, the implementation needs to consider the opponent's previous action. This is stored in the vectors *player* and *opp*, passed on as input parameters. In both strategies 3 and 4, the first check is whether the length of the opponent's vector of moves is equal to zero. If this is the case, it means that the game is in the very first round, and since both of these strategies are cooperative in nature, they start with a cooperation move, i.e., they return a value of 1 as their current play. For the rest of the game each of these two strategies plays out differently. Tit-for-Tat checks the opponent's last move, found in the last position of the vector *opp*, and plays the same by assigning the opponent's last move as the player's current move. Grim Trigger needs to check whether the opponent has defected at any time previously in the game, since if it has, according to the Grim Trigger strategy the player must forever defect. Therefore, it compares the length of the opponent's vector to the sum of the values in the opponent's vector. If the opponent does not have all values of 1 in the vector of moves, i.e., has defected and a value of 0 is included in the vector, then the sum of the values will not match the length of the vector, and this will cause the player employing the Grim Trigger strategy to defect in the current round; otherwise it will cooperate.

Table 6.4: Code in play.m

```
     function x=play(player,opp,strategy)
1.   if(strategy==1)
2.   x=0;
3.   elseif(strategy==2)
4.   x=1;
5.   elseif(strategy==3)
6.   if(length(opp)==0)
7.   x=1;
8.   else
9.   x=opp(length(opp));
10.  end
11.  elseif(strategy==4)
12.  if(length(opp)==0)
13.  x=1;
14.  else
15.  if(sum(opp) < length(opp))
16.  x=0;
17.  else
18.  x=1;
19.  end
20.  end
21.  else
22.  s=randn;
23.  if(s < 0)
24.  x=0;
25.  else
26.  x=1;
27.  end
28.  end
```

Examples of strategies discussed in Chapter 2 and Chapter 3 that are differentiated from the above-mentioned well-known strategies, are given next.

6.8   Implementing strategies with non-cooperative nature

In Chapter 2, we have presented 2 non-cooperative strategies for the Iterated Prisoner's Dilemma variation that we have named *User-Network Interaction game*. The first one is the *Cheat-and-Leave* strategy, with which the user defects in one round of the game and then leaves the interaction to avoid being punished by the network. The second one is the *Cheat-and-Return* strategy, with which the network defects in one round but not being able to leave the relationship, returns in the subsequent round of the game. Table 6.5 illustrates the implementation of the two strategies in the *play* function, specifically as additional branches in the *if-elseif-else* structure that selects between the available strategies.

The use of the randomization in the *Cheat-and-Leave* strategy, is so that it is randomly decided when to cheat. Once the randomized number is generated, we check whether it is negative and if it is the player defects, otherwise the player cooperates. This probability is 50% toward either decision, since this is the probability of generating a negative number using the function *randn*. However, once the player has defected, the player abandons the interaction and this action must be indicated by something other than 0 or 1, which indicate defection and cooperation respectively, so we choose to set it to -1 to indicate the action of *leaving*. Once the player leaves, it continues to offer that action forever, i.e., it remains away from the interaction. We define this kind of behavior both for the first move in the interaction, i.e., we allow a player to defect from the first round, as well as in the rest of the rounds in the repeated interaction.

In the *Cheat-and-Return* strategy, there is no action of *leaving*, the player does not have this kind of action available, so it can only defect and cooperate. Again, the decision to defect is decided in a random manner, similarly to the *Cheat-and-Leave* strategy with a 50% probability of defection, given by the probability that the function *randn* returns a negative value. Lines $37 - 38$ differentiates this strategy from the *Cheat-and-Leave* strategy, since it returns a 1 if the player has defected in the previous period, i.e., the player continues with cooperation even though the player has defected in the previous period.

6.9   Implementing a simple modification of the Grim Trigger strategy

In this section we present the implementation of the modified *Grim Trigger* strategy (Table 6.6), as we have proposed in Chapter 2, so that the punishment

Table 6.5: Code for non-cooperative strategies

```
1.    elseif(strategy==6)
2.    if(length(opp)==0)
3.    s=randn;
4.    if(s < 0)
5.    x=0;
6.    else
7.    x=1;
8.    end
9.    else
10.   if(player(length(player))==0 ——— player(length(player))==-1)
11.   x=-1;
12.   else
13.   s=randn;
14.   if(s < 0)
15.   x=0;
16.   else
17.   x=1;
18.   end
19.   end
20.   end
21.   elseif(strategy==7)
22.   if(length(opp)==0)
23.   s=randn;
24.   if(s < 0)
25.   x=0;
26.   else
27.   x=1;
28.   end
29.   else
30.   if(player(length(player))==1)
31.   s=randn;
32.   if(s < 0)
33.   x=0;
34.   else
35.   x=1;
36.   end
37.   else
38.   x=1;
39.   end
40.   end
```

Table 6.6: Code for Leave-and-Return strategy

1.	elseif(strategy==8)
2.	if(length(opp)==0)
3.	x=1;
4.	else
5.	if(opp(length(opp))==0)
6.	x=-1;
7.	else
8.	x=1;
9.	end
10.	end

part of this strategy is not forever but only for one round of the game. We have referred to this modification of the *Grim Trigger* strategy as the *Leave-and-Return* strategy.

The first move is always to cooperate, and this is indicated in Lines $2-3$, where a 1 is returned if the length of the vector representing the history of moves of the opponent is equal to 0. This means that the opponent hasn't played yet, therefore, the current move is the first move in the game. In the rest of the lines of this strategy implementation we see that a -1, which indicates the *leaving* action, is returned if defection is detected by the opponent in the previous round, however, the player continues with cooperation in any other case. This represents the punishment of only one round instead of forever, which is the punishment employed by the the *Grim Trigger* strategy.

6.10   Implementing adaptive strategies

A more challenging type of strategy to implement than the strategies we have seen so far, is an adaptive strategy, because it requires that the player has additional intelligence, so that it *remembers* the opponent's overall behavior and actions, We have discussed such strategies in both Chapter 2 and Chapter 3. In this section, we outline the implementation of the adaptive strategies used in Chapter 2 and Chapter 3. We illustrate the modification in the file *play.m*, where some additional elements are checked prior to deciding whether the player cooperates or defects. These include some history information, calculated outside *play.m*, which needs to be passed to this file, so the programmer should consider this additional information exchange between the files.

As we can see in Figure 6.7, the format is the same as the strategies previously described, adding the elements necessary to include the adaptivity. Line 2 indicates the cooperation action as the first move of the game. Line 5 checks whether opponent has cheated in the previous round, and in Line 6 the value α is used to calculate the number of periods, represented by variable x,

Table 6.7: Code for strategy of adaptively punishing opponent's defection (Chapter 2)

```
1.    elseif(strategy==9)
2.    if(length(opp)==0)
3.    x=1;
4.    else
5.    if(opp(length(opp))==0)
6.    x=ceil(1/alpha);
7.    elseif(opp(length(opp))==1)
8.    x=1;
9.    else
10.   if(opp(length(opp))==-1)
11.   x=1;
12.   else
13.   y=opp(length(opp));
14.   for i = 1 : y − 1
15.   if((opp(length(opp)-i)==y))
16.   temp 1;
17.   else
18.   temp=y;
19.   end
20.   end
21.   x=temp;
22.   end
23.   end
24.   end
```

that the punishment will last. Otherwise, if the opponent has cooperated in the previous period (Line7), continue cooperating (Line8). The *else* branch in Line 9 represents the punishment period and the number of periods that the punishment lasts for is represented by a negative number of the same value (to separate the number of periods of punishment from the cooperate action). In Line 10, we check if the punishment lasts for one period, which means that the opponent will not be punished in the next period, therefore Line 11 instructs the player to return to cooperation in the next period. Otherwise, the *for* loop iterates to check whether punishment is over or not, and the result, i.e., whether to currently cooperate or defect, is stored in variable x (Line 21).

Table 6.8 illustrates an implementation, based on the presented framework, of the strategy presented in Chapter 3, which adaptively selects whether the player should declare the real costs or to cheat, where declaring real costs is the cooperative behavior and cheating represents defection. Table 6.8 illustrates

Table 6.8: Code for strategy of adaptively declaring real costs (Chapter 3)

```
1.    elseif(strategy==10)
2.    if(length(player)==0)
3.    x=1;
4.    else
5.    if(beta > 0.9)
6.    x=0;
7.    y=1;
8.    else
9.    x=1;
10.   y=1;
11.   end
12.   end
```

the code addition in file *play.m* that is necessary to include this strategy in the implementation framework we have been outlining here, given that additional history information that is used in the decision needs to be passed as a parameter into this file, much like the previous implementation of an adaptive strategy illustrated in Figure 6.7.

Lines $2 - 3$ show that the first move is one of cooperation, as we have seen in many of the strategies previously described. In the few next lines of this strategy's implementation, we check the value of β, passed into this file as a parameter, which is the value that represents the probability of being truthful, and it may adaptively modify the payoffs of a player. The value of β must be kept for the whole duration of the game, so it is suggested that a placeholder for this value is defined in the *iteratedpd* function, since this is the function that calls the *play* function, which handles the play of a single round of the game. The player cheats if the value of β is high, specifically above 0.9, otherwise the player continues to declare real costs. In both cases, there is a flag, represented by variable y, set to 1, which indicates that the payoff should be adapted according to β. Moreover, the value of β itself needs to be adapted after each round of play in the file *iteratedpd.m*, which is where the function *play* is called from, because it is the return place for the decision of the player at the end of each round. Given a choice of cheating or a choice of declaring real costs, the code in *iteratedpd.m* can reflect the player's choice in the recalculation of β as this has been defined in Chapter 3.

Remark 11. *The code in this chapter includes some simple example strategies, both well-known strategies as well as more specific strategies used in particular scenarios throughout the book. We present these strategies in order to give the readers an idea of how to go about implementing an iterated strategy, and to encourage the interested reader to go ahead and expand the code with their own implementations of strategies for the Iterated Prisoner's*

Dilemma game or equivalent games, like the ones we have defined in some of the previous chapters.

References

[1] Networks Research Laboratory, University of Cyprus, *http://www.netrl.cs.ucy.ac.cy*, under *Simulators and Software*.

[2] MATLAB: The Language of Technical Computing, *The Mathworks, version 7.6.0.324, (R2008a)*, February, 2008.

[3] G. Taylor, *Iterated Prisoner's Dilemma in MATLAB*, Archive for the "Game Theory" Category, http://maths.straylight.co.uk/archives/category/game-theory, March 2007.

[4] G. Kendall, X. Yao and S. Y. Chong, *The Iterated Prisoner's Dilemma: 20 Years On*, Advances In Natural Computation Book Series, vol. 4, World Scientific Publishing Co., 2009.

[5] R. M. Axelrod, *The Evolution of Cooperation*, BASIC Books, New York, USA, 1984.

Index

Printed and bound by CPI Group (UK) Ltd, Croydon, CR0 4YY

21/10/2024

01777103-0019